Donald R. McQueen

CANADIAN NATIONAL STEAM!

Volume 4: Road Numbers 1000 to 1018 — 4-6-0 — Class F
1016 to 1178 — 4-6-0 — Class G
1200 to 1454 — 4-6-0 — Class H
1500 to 1628 — 4-6-0 — Class I

For Dianne (1942-2008)

Railfare ❋ DC Books

ABOUT THE AUTHOR

Don McQueen
[ROBERT RIEHL PHOTOGRAPHY]

Don McQueen, born in Toronto in 1938, was raised in Brockville, Ontario and received an Honour Arts degree in History from Queen's University in 1963. Moving to London, Ontario later that same year, he began a secondary school teaching career which spanned 33 years in six of the city's schools. Initially a school librarian, teaching Canadian history and geography, Don also taught and developed school curricula in the humanities, world civilizations, law, sociology, world issues, and photography.

His interest in Canadian railway history developed at an early age, although not as early as his fascination with all forms of transportation, especially trains. Brockville, both a St. Lawrence River town and a terminal for two Canadian transcontinental railways, provided a lure which was impossible to resist. His articles, background research or contributions have appeared in a number of Canadian books and publications, including *Branchline*, *CN Lines* (Canadian National Railways Historical Association), *Canadian Rail*, *Extra 2200 South*, *Kingston Rail*, *Locomotive Notes II*, Ontario Northland Railway Historical & Technical Society's *The Northlander*, and Upper Canada Railway Society's *Newsletter / Rail & Transit*. In other railway related themes, he has written illustrated articles in the *Toy Train Operating Society Bulletin* (Marx tinplate toy trains) and in *Freight Cars Journal* (a White Pass & Yukon rolling stock roster).

Constructed in Kingston is Don's premier Canadian railway history reference to date, co-authored with Bill Thomson. That volume chronicled the history, in text and image, of the Canadian Locomotive Companies of Kingston from 1854 to 1968. Now out of print, that book remains a standard source in Canadian locomotive railway history.

Living in London, another railway city, resulted in a keen interest in locomotive production at General Motors' Canadian locomotive manufacturing facility. After becoming one of the charter members in the creation of the Forest City Railway Society in 1968, he continued his commitment as one of the editors of the organization's monthly publication, *Tempo Jr.*, and created a daily electronic digest, *Froth*.

When time permitted during his career, as well as in retirement, he paced his research with another of his passions – photography. In addition to his images appearing in the above periodicals, others have been published in several books such as *Signatures in Steel*, the reissued *Narrow Gauge Railways of Canada*, *Self-Propelled Cars of the CNR* and several Morning Sun publications including *Chesapeake & Ohio in Color* and various rolling stock volumes.

In retirement, Don keeps busier than ever with all of the above pursuits, including his iris garden in London, Ontario and with his family, Heather, "Socs", Christopher and Danielle.

Book designed and typeset in Adobe Garamond Pro, ITC Garamond, and Myriad MM.
Overall book design by Ian Cranstone, Osgoode, Ontario.
Graphic grid designed by Primeau & Barey, Montreal, Quebec.
Printed and bound in Canada.

Legal Deposit, *Bibliothèque et Archives nationales du Québec* and the National Library of Canada, 3rd Trimester, 2013.

Library and Archives Canada Cataloguing in Publication
McQueen, Donald R., 1938-
  Canadian National Steam! / Donald R. McQueen.

Includes bibliographical references and indexes.
ISBN 978-1-927599-00-6 (Premiere vol. : pbk.).—ISBN 978-1-927599-01-3
Premiere vol. : bound).—ISBN 978-1-927599-02-0 (v. 2).—
ISBN 978-1-927599-03-7 (v. 3).—ISBN 978-1-927599-04-4 (v. 4).—
ISBN 978-1-927599-05-1 (v. 5).—ISBN 978-1-927599-06-8 (v. 6).—
ISBN 978-1-927599-07-5 (v. 7).—ISBN 978-1-927599-08-2 (v. 8).—
ISBN 978-1-927599-09-9 (set)

1. Steam Locomotives—Canada—History. 2. Canadian National Railways—History. I. Title.

TJ603.4.C32C34 2013   625.26'10971   C2013-900857-8

We acknowledge the financial support of the Government of Canada through the Book Publishing Industry Development Program (BPIDP) for our publishing activities.

For our publishing activities, **Railfare ❄ DC Books** gratefully acknowledges the financial support of The Canada Council for the Arts, and of SODEC.

**Canada Council   Conseil des Arts
for the Arts   du Canada**

Société
de développement
des entreprises
culturelles
Québec 🏵🏵

Railfare ❄ DC Books

Ontario office:
1880 Valley Farm Road, Unit #TP-27, Pickering, ON L1V 6B3

Business office and mailing address:
Box 666, St. Laurent Station, Montreal, QC H4L 4V9
email: railfare@videotron.ca   web: www.railfare.net

CONTENTS

Note to Readers:

All railway and builder histories and checklists are found in the Premiere Volume of *Canadian National Steam!*, Section 2, pages 52-126. Explanations for technical and specification codes or abbreviations used in this volume are found in Section 9, General Notes for the Roster Project on pages 250-256.

F CLASS: 1000-1018

F-1-a
F-1-b
F-1-c
F-2-a
F-3-a
F-4-a
(F-5-A)

CNR 1000-1018

F CLASS
4-6-0 TEN WHEEL TYPE

The "F" class was assigned road numbers **1000-1015** for Ten Wheelers with drivers less than 52 inches. The class originally totalled seventeen and came from five sources: thirteen from the **Canadian Government Railway**; one from the **Canadian Northern Railway**; two from the **Quebec Oriental Railway**; and one from **Atlantic Quebec & Western Railway**. Then, in 1950, an additional five came from the **Temiscouata Railway**, raising the total to twenty-two. In 1919, the road number allotment assigned **1000-1012** and **1013** to thirteen former CGR and one CaNoR 4-6-0s respectively. In 1929, 2nd **1013** and 1st **1014** were used for the QOR, and 1015 for the **AQ&W**. In 1950, 2nd **1014-1018** were assigned to the five from the TMR.

Although the distribution of the F class sub-series assigned by the CNR Motive Power Department in 1919 may appear to be lacking sequential order, it should be noted the class was built around CaNoR classes still in use. CaNoR vacancies were reused to classify CGR and GTR motive power both in 1919 and 1923 respectively. The practice continued for other stock acquired after 1924. These reassigned CaNoR classes are listed in Part Two: *CNR Class Summary*. All dates are mm-dd-yr.

FIGURE FG
Multiple Use of Road Numbers 1000-1018 Assigned to 4-6-0s (1912-1950)

Rd. No.	First Use	Owner		Builder	Serial	Date	First Owner		Later Owners	Scrapped
1000	1-1912	CaNoR	F-2-A	Pit	#728	4-1884	BCR&N	79	BCR&N 143; CRI&P 1106; CaNoR 141	So 6-1914
	9-1919	**CNR**	**F-1-a**	MLW	#46203	4-1909	INBR	4	CGR 1153	9-1950
1001	1-1912	IRC–CGR		CLC	#437	6-1892	IRC	101:2nd		12-1917
	1-1912	CaNoR	F-2-A	Pit	#747	10-1884	BCR&N	85	BCR&N 149; CRI&P 1111; CaNoR 142	11-1916
	9-1919	**CNR**	**F-1-a**	MLW	#49469	12-1910	INBR	5	CGR 1154	8-1939
1002	1-1912	IRC–CGR		CLC	#397	12-1890	IRC	204	—	So 12-1917
	1-1912	CaNoR	F-2-B	Bro	#1082	10-1885	BCR&N	90	BCR&N 164; CRI&P 1116; CaNoR 145	9-1917
	9-1919	**CNR**	**F-1-b**	MLW	#49898	4-1911	OM&O	8	CGR 4522	11-1950
1003	1-1912	IRC–CGR		CLC	#398	12-1890	IRC	205	—	So 12-1917
	1-1912	CaNoR	F-2-C	Pit	#1194	10-1891	T&OC	58	T&OC 365; CaNoR 95	7-1917
	9-1919	**CNR**	**F-1-b**	MLW	#49896	4-1911	OM&O	12	OM&O 9; CGR 4523	9-1940
1004	1-1912	IRC–CGR		CLC	#399	12-1890	IRC	206		12-1917
	1-1912	CaNoR	F-2-D	Pit	#1196	3-1891	B&O	1500	CaNoR 167	11-1916
	9-1919	**CNR**	**F-1-b**	MLW	#51127	4-1912	OM&O	10	CGR 4524	12-1954
1005	1-1912	IRC–CGR		CLC	#419	10-1891	IRC	207	**CNR G-18-a 1166**:1st	2-1922
	1-1912	CaNoR	(F-2-E)	Pit	#745	9-1884	BCR&N	83	BCR&N 147; CRI&P 1109; CaNoR 149	-1911?
	9-1919	**CNR**	**F-1-b**	MLW	#51128	4-1912	OM&O	11	CGR 4525	6-1936
1006	1-1912	IRC–CGR		CLC	#420	10-1890	IRC	208	—	So 4-1918
	1-1912	CaNoR	(F-2-E)	Pit	#746	10-1884	BCR&N	84	BCR&N 148; CRI&P 1110; CaNoR 168	6-1911
	9-1919	**CNR**	**F-1-b**	MLW	#51129	4-1912	OM&O	12	CGR 4526	6-1936
1007	1-1912	IRC–CGR		CLC	#421	10-1891	IRC	209	—	12-1917
	1-1912	CaNoR	F-3-A	Pit	#—	-1893	nd		JBR 161	11-1915
	9-1919	**CNR**	**F-1-b**	MLW	#51130	4-1912	OM&O	13	CGR 4527	8-1948
1008	1-1912	IRC–CGR		CLC	#422	10-1891	IRC	210	—	12-1917
	1-1912	CaNoR	F-3-A	Pit	#—	-1893	nd		JBR 162	6-1917
	9-1919	**CNR**	**F-1-b**	MLW	#51131	5-1912	OM&O	14	CGR 4528	8-1954
1009	1-1912	IRC–CGR		CLC	#433	3-1894	IRC	10:2nd	—	So 12-1918
	1-1912	CaNoR	F-3-A	Pit	#—	-1893	nd		JBR 163	7-1917
	9-1919	**CNR**	**F-1-b**	MLW	#51132	5-1912	OM&O	15	CGR 4549; **CNR F-1-b 1165**:2nd	Dn 6-1960
1010	1-1912	IRC–CGR		CLC	#434	4-1894	IRC	11:3rd	—	12-1917
	1-1912	CaNoR	F-4-A	RI	#2543	6-1891	TStL&KC	54	TStL&W 54; CaNoR 93; **CNR F-4-a 1013**:1st	12-1920
	9-1919	**CNR**	**F-1-b**	MLW	#51133	5-1912	OM&O	16	CGR 4530	7-1926
1011	1-1912	IRC–CGR		CLC	#435	4-1894	IRC	12:4th	—	12-1917
	1-1912	CaNoR	F-4-A	RI	#2196	2-1890	TStL&KC	47	TStL&W 47, 82; CaNoR 94; **CNR F-4-a 1014**	12-1920
	9-1919	**CNR**	**F-1-b**	MLW	#49919	5-1911	FCC	7	OF&M 17; CGR 4531	5-1939
1012	1-1912	IRC–CGR		CLC	#436	4-1894	IRC	14:3rd	—	12-1917
	1-1912	CaNoR	F-5-A	PRR	#????	?-18??	PRR	104?	CaNoR 96	3-1918
	9-1919	**CNR**	**F-2-a**	PITTS	#38877	12-1905	NWGR	4	OF&M 20; CGR 4501; **CNR F-2-a 1166**:2nd	3-1958
1013	(1-1912)	CaNoR-	F-5-A	PRR	#1208?	-1893?	PRR	??	CaNoR 169	Wr 12-1911
	1-1912	CaNoR	G-1-B	??	Rblt StLL	-1911	(USA) ?	47	**CNR G-1-b 1016**	10-1923
	9-1919	**CNR**	**F-4-a**	RI	#2543	6-1891	TStL&KC	54	TStL&W 54; CaNoR 93, CaNoR 1010	12-1920
	10-1929	**CNR**	**F-3-a**	MLW	#46562	9-1909	NCCo	4	QOR 14	6-1931
1014	1-1912	CaNoR	F-5-A	PRR	#1209?	-1893?	PRR	??	CaNoR 170	8-1918
	(9-1919)	**CNR**	**F-4-a**	RI	#2196	2-1890	TStL&KC	47	TStL&W 47, 82; CaNoR 94; CaNoR 1011	4-1917
	10-1929	**CNR**	**F-3-a**	MLW	#46563	9-1909	NCCo	5	QOR 15	8-1931
	1-1950	**CNR**	**F-1-c**	MLW	#50132	4-1911	NBCR	6	TMR 6	4-1956
1015	1-1912	CaNoR	G-1-A	Brooks	#1079	9-1895	BCRN	92	BCRN 151; CRI&P 1213, 1113; CaNoR 144	11-1916
	(9-1919)	**CNR**	**F-5-a**	PRR	#????	?-18??	PRR	104?	CaNoR 96, CaNoR 1012	3-1918
	10-1929	**CNR**	**F-4-a**	MLW	#49920	5-1911	FCC	8	OM&M 18; CGR 4532; AQ&W 33	6-1931
	1-1950	**CNR**	**F-1-c**	MLW	#49908	5-1911	C&PC	7	TMR 7	12-1954
1016	1-1912	CaNoR	G-1-A	Pitts	#729	4-1884	BCRN	80	BCRN 144; CRI&P 1218, 1118; CaNoR 147	12-1916
	9-1919	**CNR**	**G-1-b**	??	Rblt StLL	0-1911	(USA) ?	607	CaNoR 1013	10-1923
	1-1950	**CNR**	**F-1-c**	MLW	#46205	9-1911	TMR	8	—	3-1955
1017	1-1912	CaNoR	G-2-A	RI	#2392	6-1890	ATSF	463	ATSF 509, 145; CaNoR 63	4-1920
	9-1919	**CNR**	**G-2-a**	RI	#2392	6-1890	ATSF	463	ATSF 509, 145; CaNoR 63, CaNoR 1017	4-1920
	1-1950	**CNR**	**F-1-c**	MLW	#46206	9-1911	TMR	9	**CNR F-1-c 1167**:2nd	7-1958
1018	1-1912	CaNoR	G-2-A	RI	#2411	6-1890	ATSF	467	ATSF 507, 149; CaNoR 64	6-1921
	9-1919	**CNR**	**G-2-a**	RI	#2411	6-1890	ATSF	467	ATSF 507, 149; CaNoR 64, CaNoR 1018	6-1921
	1-1950	**CNR**	**F-1-c**	MLW	#49897	4-1911	OM&O	7	CGR 4521; TMR 10; **CNR F-1-c 1168**:2nd	5-1959

CNR 1000

4-6-0 TEN WHEEL TYPE

F-1-a

Cylinder	Gear	Driv.	Pressure	Boiler	T.E.	Haulage	Steam	Stkr.	Drivers/Eng./Total	Water	Coal	Length	Notes
			Specifications				Appliances		Weights	Fuel Capacity		Length	Notes
18x24"	S	51"	160#	EWT	20736		sat		84/110/203380	3500 gals	8 tons	57-6'	[MLW]
18x24"	S	51"	160#	EWT	20736	21%	sat		85/111/203000	3500 gals	8 tons	57-6'	[CNR]

	Serial	Built	New as	8-1914	Stl Cab	Disposition	To
Montreal Locomotive Works Ltd. – ALCO		1911	(Q-107)			(1) Acquired by CNR 9-01-1919	
			—	T1-8 105%			
1000	46203	4- -09	**INBR 4**	**CGR 1153**	3-24 AV	Sc 9-09-50 AK	
—	46205	4- -09	TMR 8*				**CNR 1016**/2
—	46206	4- -09	TMR 9				**CNR 1017**/2

CNR 1000 was built for the **International Railway of New Brunswick** along with **Temiscouata Railway** 8 and 9 (which later became **CNR** second **1016-1017** in 1950), likely as stock. They were all shipped in September 1909. The MLW record does not show serial number 46204 applied to any locomotive, raising the possibility one of the customers subsequently reduced their initial order.

Although taken onto CNR books thirty-two years apart, the three locomotives built as stock in MLW order Q-107 were purchased by two owners. As-built **TMR 8**, at Montreal in September 1911,
[MLW PHOTO Q-107/SIRMAN COLLECTION]
became CNR second 1016 in 1950 (see page F-9), whereas **1000**, at Stellarton, on September 1st 1941, had been acquired in 1919.
[AL PATERSON COLLECTION]

CNR 1001

4-6-0 TEN WHEEL TYPE

F-1-a

Cylinder	Gear	Driv.	Pressure	Boiler	T.E.	Haulage	Steam	Stkr.	Drivers/Eng./Total	Water	Coal	Length	Notes
			Specifications				Appliances		Weights	Fuel Capacity		Length	Notes
18x24"	S	51"	160#	EWT	20736		sat		85/111/199500	3500 gals	8 tons	57-6'	[CRMW]
18x24"	S	51"	160#	EWT	20736	21%	sat		85/111/203000	3500 gals	8 tons	57-6'	[CNR]

	Serial	Built	New as	8-1914		Disposition
Montreal Locomotive Works Ltd. – ALCO		1910	(Q-148)		(1) Acquired by CNR 9-01-1919	
			—	T1-8 105%		
1001	49469	12- -10	**INBR 5**	**CGR 1154**		Sc 8-31-39 AK

CNR 1001 was built for the **International Railway of New Brunswick**. *CNR Mechanical Department Locomotive Diagrams* before 1950 show **CNR 1001** with serial number #46205 (see **CNR** second **1016** on page F-8) and a build-date of 1909, but makes no reference to the MLW order number Q-148. The MLW record shows the 1910 date used in this roster. By the time **CNR** second **1016**'s data was added to Issue Q of the redrawn *CNR Mechanical Department Locomotive Diagrams* in 1950 under #46205, **CNR 1001** and its serial number had been removed from the sheet.

In the scrap line at Moncton in 1939, **1001** had been the only locomotive built under order Q-148, but to the same specifications as those in Q-107.
[*CNR LOCOMOTIVE DATA CARD*]
Very few structural changes to 1000 (page F-3) and 1001 had taken place throughout their service years, although pilots, pony trucks and piping had invariably been altered. The steel cab retrofitted to the 1000 was one significant difference between it and the 1001.

CNR 1002-1003 — 4-6-0 TEN WHEEL TYPE — F-1-b

Cylinder	Gear	Driv.	Pressure	Boiler	T.E.	Haulage	Steam	Stkr.	Drivers/Eng./Total	Water	Coal	Length	Notes
			Specifications				Appliances		Weights	Fuel Capacity		Length	Notes
18x24"	S	50"	160#	EWT	21150		sat		84/109/196500	4000 gals	8 tons	57-6'	
18x24"	S	51"	160#	EWT	20736	21%	sat		84/109/196500	4000 gals	8 tons	57-6'	

Montreal Locomotive Works Ltd. – ALCO 1911 (Q-161) $11,800 (2) Acquired by CNR 9-01-1919

	Serial	Shipped	New as	4-1912	nd	3-1916	9-05-1918	11-29-1918	Stl cab	Disposition	To
			—	—		T1-8^ 105%					
1003	49896	4- -11	OM&O 12	OM&O 9		CGR 4523			1-25 AV	Sc 9-26-40 AK	
—	49897	4- -11	OM&O 7			CGR 4521	TMR 10				CNR 1018/2
1002	49898	4- -11	OM&O 8			CGR 4522			5-47 AK	Sc 11-15-50 AK	
—	49908	5- -11	C&PCo 7	—	TMR 7						CNR 1015/2
—	49919	5- -11	FCC 7		OF&M 17	CGR 4531					CNR 1011
—	49920	5- -11	FCC 8		OM&O 18	CGR 4532		AQ&W 33			CNR 1015/1

CNR 1002 and **1003** were from a MLW stock order for six 4-6-0s, completed in April 1909, which were sold in 1911 to three National Transcontinental Railway contractors: **O'Brien, McDougall & O'Gorman; Cavicchi & Pagano** and **E. F. & G. E. Fauquier (Construction Company)**. All six eventually became CNR F class locomotives but via different routes. A pair came close to being assigned the same CNR road number, namely **1015**.

The disposition of **1002**'s tender, not scrapped with the engine at Moncton in 1950, is not known.

OM&O 12 was renumbered to OM&O 9 to clear the series for the next order of 4-6-0s (see **CNR 1004-1010** on page F-5). When the **National Transcontinental Railway** contracts were completed, the OM&O locomotives were sold to the **Canadian Government Railways**. Two of the three became **CNR 1002-1003**, the other, OM&O 7 (as CGR 4521), was sold to the **Temiscouata Railway**. It eventually became **CNR** second **1018** in 1951. During its TMR career **1018** was modified with Economy steam chests (EsC) and outside steam pipes, but never acquired superheaters.

C&PCo 7 was sold to the **Temiscouata Railway**, later becoming **CNR** second **1015** in 1950.

Fauquier 7-8 were sold to the OM&O when the former's **National Transcontinental Railway** contracts were finished. They, as were the other OM&O 4-6-0s in this lot, were sold to the **Canadian Government Railways**. OM&O 17 (as CGR 4531) became **CNR 1011**; OM&O 18 (as CGR 4532) was sold to the **Atlantic Quebec & Western Railroad**, and didn't join the CNR roster until 1929. It was assigned road number first **1015** but was scrapped as AQ&W 33.

Records before 1924, including the *CNR Locomotive Data Card* entries of 1-01-1924, show driver diameters as 50 inches, but subsequent *CNR Locomotive Diagram* sheets show diameters as 51 inches.

CNR 1004-1010

4-6-0 TEN WHEEL TYPE

F-1-b

			Specifications				Appliances		Weights	Fuel Capacity		Length	Notes
Cylinder	Gear	Driv.	Pressure	Boiler	T.E.	Haulage	Steam	Stkr.	Drivers/Eng./Total	Water	Coal		
18x24"	S	50"	160#	EWT	21150		sat		84/109/196500	4000 gals	8 tons	57-6'	
18x24"	S	51"	160#	EWT	20736	21%	sat		84/109/196500	4000 gals	8 tons	57-6'	

Montreal Locomotive Works Ltd. – ALCO			1912	(Q-195)	$11,800						(7) Acquired by CNR 9-01-1919		
	Serial	Shipped	New as	3-1916			Stl. cab	Mod	Tender to	10-13-1957		Disposition	To
			—	T1-8ᴬ	105%					to F-1-b			
1004	51127	4- -12	OM&O 10	CGR 4524			-nd					Sc 12-07-54 A	
1005	51128	4- -12	OM&O 11	CGR 4525			-nd	f				Sc 6-08-36 AK	
1006	51129	4- -12	OM&O 12	CGR 4526								Sc 6-08-36 AK	
1007	51130	4- -12	OM&O 13	CGR 4527			-nd		CN 52268			Sc 8-21-48 AK	
1008	51131	5- -12	OM&O 14	CGR 4528			c-38					Sc 8-16-54 A	
1009	51132	5- -12	OM&O 15	CGR 4529			7-32 AV	m?		CNR 1165/2	m	Dn 6-23-60 C	CRHA
1010	51133	5- -12	OM&O 16	CGR 4530	Wr 7-30-26							Sc 7-30-26 A	

CNR 1004-1010 were built for **O'Brien, McDougall & O'Gorman**, and after the completion of the **National Transcontinental Railway** they were sold to the **Canadian Government Railways**. In 1960, CNR 1009 (as **CNR 1165**) was donated to the **Canadian Railroad Historical Association** to become part of the **Canadian Railway Museum**. In 1983, the CRM loaned the 4-6-0 to affiliate CRHA–New Brunswick Division to operate as **CNR 1009** on the Salem & Hillsborough Railroad. **CNR 1010** was destroyed in a wreck in the Atlantic Region in 1926.

The disposition of the tenders for **1005** and **1006**, not scrapped with the engines at Moncton in 1936, are not known. The tender of the **1007**, renumbered to **CN 52268**, was assigned Atlantic Region OCS between 1948 and its scrapping in November 1961.

Records before 1924, including the *CNR Locomotive Data Card* entries of 1-01-1924, show driver diameters

Even though the thirteen Ten Wheelers built under MLW Q-161 and Q-195 were for three NTR contractors, all eventually became members of the F-1-b class. **CGR 4524** (1004) was at an unidentified location, but likely on the NTR between 1916 and 1919.
[H.L. GOLDSMITH/GEORGE CARPENTER COLLECTION]
Although badly cropped, this image was one of very few photographs taken of the 4-6-0s during the construction of Canada's second transcontinental line. Unlike CNR 1002 (see Vol. 1, p. 98.) which remained relatively unaltered from its construction days, 1008, at Kent Jct., New Brunswick on July 1st 1940, had been "modernized" with a steel cab and relocated headlight.
[AL PATERSON COLLECTION]

as 50 inches, but subsequent *CNR Locomotive Diagram* sheets show diameters as 51 inches.

See Figure FG (page F-2) for a summary of the multiple use of 4-6-0 road numbers **1010-1018** between the years 1912 and 1950.

F-1-b

F-2-a

F-4-a

CNR 1011 — 4-6-0 TEN WHEEL TYPE — F-1-b

Cylinder	Gear	Driv.	Pressure	Boiler	T.E.	Haulage	Steam	Stkr.	Drivers/Eng./Total	Water	Coal	Length	Notes
			Specifications				Appliances		Weights	Fuel Capacity		Length	Notes
18x24"	S	50"	160#	EWT	21150		sat		84/109/196500	4000 gals	8 tons	57-6'	
18x24"	S	51"	160#	EWT	20736	21%	sat		84/109/196500	4000 gals	8 tons	57-6'	

Montreal Locomotive Works Ltd. – ALCO	1911	(Q-161)	$11,800				(1) Acquired by CNR 9-01-1919
	Serial	Shipped	New as	5-1912?	3-1916	Stl. cab	Disposition
			—		T1-8^A 105%		
1011	49919	5- -11	**FCC 7**	**OF&M 17**	**CGR 4531**	-nd	Sc 5-31-39 AK

CNR 1011. See 1002-1003 (page F-4).

CNR 1012 — 4-6-0 TEN WHEEL TYPE — F-2-a

Cylinder	Gear	Driv.	Pressure	Boiler	T.E.	Haulage	Steam	Stkr.	Drivers/Eng./Total	Water	Coal	Length	Notes
			Specifications				Appliances		Weights	Fuel Capacity		Length	Notes
18x24"	S	50"	160#	EWT	21120		sat		83/107/190550	3400 gals	5 tons	56-6'	[1917]
18x24"	S	51"	160#	EWT	20736	21%	sat		83/107/190550	3400 gals	5 tons	56-6'	[CNR]
18x24"	S	51"	160#	EWT	20736	21%	sat		83/107/198400	3500 gals	9 tons	60-1½'	[sb tender]

Pittsburgh Locomotive & Car Works – ALCO	1905	(P-494)	$10,884							(1) Acquired by CNR 9-01-1919
	Serial	Shipped	New as	3-1909	?-1912?	3-1916	Stl. cab	Mod	5-1952	9-13-1957	Disposition
						T1-8^A 105%			Tender of	F-2-a	
1012	38877	12-30-05	**NNWR 4**	**(FDDM&S)**	**OM&O 20**	**CGR 4501**	3-35	m	**CNR 7314**	**CNR 1166**/2	Sc 3-21-58 AK

CNR 1012 was built for the **Newton & Northwestern Railroad** before it was absorbed into the **Fort Dodge, Des Moines & Southern Railroad** in 1909. It was subsequently sold by the interurban line to **O'Brien, McDougall & O'Gorman** for construction work on the **National Transcontinental Railway**. When acquired by the **Canadian Government Railways**, it remained assigned to the NTR.

CNR 1012, posed with its crew at Truro, Nova Scotia about 1954, was a one-of-a-kind for several reasons. Built by Pittsburgh, it was the only member of the F class to see service on a regional road in the USA, and known to have acquired a slope-backed switching tender. [AL PATERSON COLLECTION]

In May 1952, the 4-6-0 received a slope-backed 0-6-0 tender from CNR O-15-c 7314.

CNR 1013-1014 (first) — 4-6-0 TEN WHEEL TYPE — first F-4-a

Cylinder	Gear	Driv.	Pressure	Boiler	T.E.	Haulage	Steam	Stkr.	Drivers/Eng./Total	Water	Coal	Length	Notes
			Specifications				Appliances		Weights	Fuel Capacity		Length	Notes
18x24"	S	55"	160#		19200		sat		81/108/ 000	gals	tons	- '	[TSL&KC 47]
18x24"	S	54"	160#		19600		sat		81/108/ 000	gals	tons	- '	[TSL&KC 54]
18x24"	S	52½"	150#		18884	18%	sat		81/108/ 000	4700 gals	tons	- '	[CaNoR 1903]
18x24"	S	50"	150#		16000	16%	sat		81/107/ 000	gals	tons	- '	[CNR]

Rhode Island Locomotive Company	1890; 1891						(1) Acquired by CNR 9-01-1919
	Serial	Shipped	New as	1901	1907	1-1912	Disposition
			—	(1904 Class)	—	F-4-A	
1013/1	2543	6- -91	**TSL&KC 54**	**TSL&W 54** Ea	(JTG) CaNoR 93	5-23-07 **CaNoR 1010**	Sc 3-21-20 AK
(1014)/1	2196	2- -90	TSL&KC 47	TSL&W 82 E	(JTG) CaNoR 94	6-05-07 CaNoR 1011	Rs by 4-30-17 AK

CNR 1013 first, (1014) and G-2-b 1021-1024 were six 4-6-0s purchased through the dealer **James T. Gardiner** from the **Toledo, St. Louis & Western Railroad** in 1907. They had been built for the TStL&W's predecessor, the **Toledo, St. Louis & Kansas City Railroad**. Although CaNoR 1011 (1014) had been removed from service, it apparently was still in stock, as it was assigned a CNR road number. See the note under **CNR G-2-b 1021-1024** for more details (page G-6).

CNR 1013-1014 (second)						4-6-0 TEN WHEEL TYPE					first F-3-a
Specifications						Appliances		Weights	Fuel Capacity		Length	Notes
Cylinder	Gear	Driv.	Pressure	Boiler	T.E.	Haulage	Steam	Stkr.	Drivers/Eng./Total	Water	Coal	
18x24"	S	51"	165#		00	21%	sat		83/111/197500	3500 gals	8 tons	- '	[orig. CRMW]

Montreal Locomotive Works Ltd. – ALCO		1909	(Q-114)				(2) Acquired by CNR 10-01-1929
	Serial	Shipped	New as	-1909	-1916		Disposition
			—	—			
(1013)/2	46562	9- -09	NCCo 4	AQ&W 4?	QOR 14		Sc 6-30-31 AK
(1014)/2	46563	9- -09	NCCo 5		QOR 15		Sc 8-27-31 AK

Two examples of MLW-built 4-6-0s from two lines absorbed by CNR in 1929 didn't survive long enough to be relettered, although they had been assigned CNR classes and road numbers. Both **QOR 14** (1013) and **AQ&W 33** (1015) were photographed in the Moncton scrap line in 1931. Although with slightly different specifications, the pair of saturated Ten Wheelers had retained most of their as-built features, particularly the horizontal slatted pilots, round number plates, location of headlights, and wooden cabs. Only AQ&W 33 appeared to have had its tender modified with steel sides.
[BOTH:
CNR LOCOMOTIVE DATA CARD]

F-3-a

F-4-a

F-1-c

F-4-a

(F-5-A)

CNR **1013** and **1014** second, were built for the **New Canadian Company** and reported by the trade press for use on the **Atlantic Quebec & Western**. It was later transferred to the **Quebec Oriental Railway**.

1015						4-6-0 TEN WHEEL TYPE							(F-5-A)
			Specifications				Appliances		Weights	Fuel Capacity		Length	Notes
Cylinder	Gear	Driv.	Pressure	Boiler	T.E.	Haulage	Steam	Stkr.	Drivers/Eng./Total	Water	Coal		
18x22"	S	51"	145#		18296	18%	sat		82/104/171300	3500 gals	tons	-'	[CaNoR 96]
18x24"	S	50"	145#	BEL	17285	17%	sat		70/ 90/ 000	3500 gals	tons	-'	[CaNoR 169-170]

Pennsylvania Railroad – Indianapolis	18??	(CaNoR 96)
Pennsylvania Railroad – Indianapolis	1893?	(CaNoR 169-170)

	Serial	Shipped	New as		1907		Delivered	1-1912		Disposition
			—		—			F-5-A		
(1015)	????	-???	PRR 104 ?	(JTG)	CaNoR 96	5-30-07	6-12-07	CaNoR 1012		Sc 3-31-18 GV
—	1208?	-93?	PRR ??	(JTG)	CaNoR 169	2-01-07	2-16-07	CaNoR (1013)/1	Wr -11	Sc 12-11-11
—	1209?	-93?	PRR ??	(JTG)	CaNoR 170	2-08-07	2-12-07	CaNoR 1014		Sc 8-03-18

CNR 1015 (CaNoR 1012-1014) were purchased second-hand by the **Canadian Northern Railway** in 1907 from **James T. Gardiner** of Chicago. The dealer's records apparently gave their ancestry as being from the **Pennsylvania Railroad** – although no substantiating evidence has been found in PRR records.

The PRR road number (104) for CaNoR 1012 could be a serial number or a typographical error, for there is no record of a PRR 104 as a 4-6-0 – although there was a 4-6-0 with the number 404. CaNoR 1012 was assigned **CNR number 1015** but was listed as scrapped in 1918 – even though its tender had been sold to **William Gartshore** in October 1914, and the engine towed to Moncton for final dismantling in 1920.

CaNoR 1013 (first) and 1014 were listed as built under serials #1208 and #1209. However PRR records show serial #1208 assigned a class H-3 2-8-0 built at Altoona in 1887, and #1209 to a 0-6-0 at Juniata in 1906. Thus the build-date of 1893 may be suspect: they could have both been built in 1887 and rebuilt in 1893, for no locomotive built in 1905 would have been offered for resale in 1907.

Even the demise of CaNoR 169 (first 1013) is unclear. The *CaNoR Stock Book* shows it "destroyed in wreck MP 289½ on Can. Nor. Ontario, Dec. 11, 1911". Even though it had been assigned by CaNoR to F-5-A 1013 in the 1912 reorganization plan, CaNoR scrap record shows CaNoR old 169 "Scrapped Dec. 1917 – destroyed in wreck". Either six years passed between the wreck and dismantling, or most likely, the scrap record has a typographical error for the last digit of the date (169's entry has a ditto mark in a group of surrounding entries given as "Dec. 1917"), particularly when the old road number was used, and second CaNoR 1013 had been on the roster since 1912.

CNR 1015 (first)						4-6-0 TEN WHEEL TYPE							second F-4-a
			Specifications				Appliances		Weights	Fuel Capacity		Length	Notes
Cylinder	Gear	Driv.	Pressure	Boiler	T.E.	Haulage	Steam	Stkr.	Drivers/Eng./Total	Water	Coal		
18x24"	S	50"	160#	EWT	21150		sat		84/109/196500	4000 gals	8 tons	57-6'	[CRMW]
18x24"	S	51"	165#	EWT	20736	21%	sat		/108/ 000	gals	tons	55-8'	

Montreal Locomotive Works Ltd. – ALCO	1911	(Q-161)	$11,800			(1) Acquired by CNR 10-01-1929
	Serial	Shipped	New as	?-1916?	3-1916	11-29-1918	Disposition
			—	—	T1-8^A 105%		
(1015)/1	49920	5- -11	FCC 8	OM&O 18	CGR 4532	AQ&W 33	Sc 6-30-31 AK

CNR 1015 first. See **1002-1003** (page F-4).

CNR 1014-1017 (third; second)						4-6-0 TEN WHEEL TYPE							F-1-c
			Specifications				Appliances		Weights	Fuel Capacity		Length	Notes
Cylinder	Gear	Driv.	Pressure	Boiler	T.E.	Haulage	Steam	Stkr.	Drivers/Eng./Total	Water	Coal		
18x24"	S	51"	160#	EWT	20735		sat		84/110/203380	3500 gals	8 tons	57-6'	[MLW, TMR 8-9]
18x24"	S	50"	160#	EWT	20735		sat		84/109/196500	4000 gals	8 tons	57-6'	[CRMW 1015]
18x24"	S	51"	160#	EWT	20735	21%	EsC		95/118/ 000	4000 gals	8 tons	57-6'	

Montreal Locomotive Works Ltd. – ALCO	1909 & 1911	(Q-107, Q-161 & Q-166)	$11,800						(4) Acquired by CNR 1-01-1950
Serial		Built	Shipped	New as	10-1914?	1916?	To CNR	Mod	9-13-1915	Tender	Disposition
				—	—	—	F-1-c		F-1-c BA	to	
1014/3	50132	Q-166	4- -11	11- -11	NBCR 6	CECo 6	TMR 6	1-51	m?		OCS	Sc 4-18-56 AK
1015/2	49908	Q-161	4- -09	5- -11	C&PCo 7	—	TMR 7	12-50	m			Sc 12-31-54 AK
1016/2	46205	Q-107	4- -09	9- -09	TMR 8*			12-50	m			Sc 3-02-55 AK
1017/2	46206	Q-107	4- -09	9- -09	TMR 9			12-50	m?		CNR 1167/2	Sc 7-21-58 AK

CNR 1014 third, was built under MLW order Q-166, likely as stock, and was later purchased by the **New Brunswick Coal & Railway Company**. It worked between Norton and Chipman at least until the CPR took over operations between Norton and Minto on October 18th 1914. Sometime later, it was sold to the **Temiscouata Railway** through the dealer **Canadian Equipment Company**. The tender of **1014** was listed on December 31st 1955 as held for possible conversion to an OCS assignment, but no other record has been found.

During their TMR careers, **1014-1017** were modified with Economy steam chests (EsC) and outside steam pipes, but never superheated.

The five locomotives which became the F-1-c class were to be the third-last steam acquisition made by CNR. **TMR 8**, at Edmundston, New Brunswick on September 15th 1943, had been altered from its appearance when delivered (see TMR 8, page F-3) by superheating with outside steam pipes, and by changes made in type of pilot, location of handrails and air reservoir tanks, the addition of a steel cab, and different lettering and numbering locations. [FRED J. SANKOFF PHOTO/SIRMAN COLLECTION]

CNR 1016 (TMR 8), at Moncton on July 30th 1951, had by that date been completely transformed to CNR standards, with an off-centred headlight and a road number lamp uncharacteristically mounted above the top of the smokebox. [PETER COX/ROBERT SANDUSKY COLLECTION]

CNR 1015 second. See **CNR 1002-1003** (page F-4).
CNR 1016-1017 second. See **CNR 1000** (page F-3).

CNR 1018 (second) 4-6-0 TEN WHEEL TYPE F-1-c

Cylinder	Gear	Driv.	Specifications Pressure	Boiler	T.E.	Haulage	Appliances Steam	Stkr.	Weights Drivers/Eng./Total	Fuel Capacity Water	Coal	Length	Notes
18x24"	S	51"	160#	EWT	20735	21%	EsC		95/118/ 000	4000 gals	8 tons	57-6'	

Montreal Locomotive Works Ltd. – ALCO			1911	(Q-161)	$11,800				(1) Acquired by CNR 1-01-1950
	Serial	Built	Shipped	New as	3-1916	9-05-1918	To CNR	Mod	9-13-1957 BA	Disposition
				—	T1-8ᴬ—105%	—	F-1-c		F-1-c	
1018/2	49897	4- -09	4- -11	OM&O 7	CGR 4521	TMR 10	1-51	m	CNR 1168/2	Sc 5-15-59 AK

CNR 1018 second. See 1002-1003 (page F-4).

CNR 1018, at Rivière-du-Loup after August 1952, had also lost any resemblance to its appearance as a Temiscouata steamer. To improve crew safety, Moncton shops had raised and extended the running boards on both it and 1016 to clear the steam pipes and provide ladders to the pilot deck. The longitudinal air reservoir tank was relocated under the running boards. [AL PATERSON COLLECTION]

FIGURE FC
Multiple Use of Similar 4-6-0 F Classes by CaNoR and CNR (1912-1957)

Class	CaNoR			CNR			
(F-1-a)	27	Sc. 1911					
	28	Sc. 1911					
F-1-a				1000	1919-1950		
				1001	1919-1939		
(F-1-b)	29	Sc. 1911					
F-1-b				1002	1919-1950		
				1003	1919-1940		
				1004	1919-1954		
				1005	1919-1936		
				1006	1919-1936		
				1007	1919-1948		
				1008	1919-1954		
				1009	1919-1957;	as 1165/2	1957-1960
				1010	1919-1926		
				1011	1919-1939		
(F-1-c)	30	Sc. 1911					
F-1-c	30			1014/3	1950-1956		
				1015/2	1950-1954		
				1016/2	1950-1955		
				1017/2	1950-1957;	as 1167/2	1957-1958
				1018/2	1950-1957;	as 1168/2	1957-1959
F-2-a	1000	1912-1914					
	1001	1912-1916					
F-2-a				1012	1919-1957;	as 1166/2	1957-1958
F-2-b	1002	1912-1917					
F-2-c	1003	1912-1917					
F-2-d	1004	1912-1916					
F-2-e	(1005)	Sc. 1911?					
	(1006)	Sc. 1911					
F-3-a	1007	1912-1915					
	1008	1912-1917					
	1009	1912-1917					
F-3-a				1013/2	1929-1931		
				1014/2	1929-1931		
F-4-a	1010	1912-1919;	as	1013/1	1919-1920		
	1011	1912-1917;	as	(1014/1	1917)		
F-5-a	1012	1912-1918	as	(1015/2	1918)		
	(1013/1)	Wr. 1911					
	1014	1912-1918					

CONTENTS

G CLASS: 1016-1178

(G-1-A)
G-15-a
G-1-b
G-16-a
G-2-a
G-17-a
(G-2-B)
G-18-a
G-2-c
G-19-a
G-3-a
G-20-a
G-4-a
G-21-a
(G-5-A)
G-5-a
G-5-b
G-6-a
G-7-a
G-8-a
G-9-a
G-10-a
G-10-b
G-11-a
G-12-a
G-13-a
G-14-a

CNR 1016-1178

G CLASS
4-6-0 TEN WHEEL TYPE

The "G" class was assigned road numbers **1016-1199** for Ten Wheelers with drivers between 52 and 58 inches. The class totalled 157 locomotives which came from five lines: 144 from **Canadian Northern Railway**, eleven from the **Grand Trunk Railway of Canada** (including one from GTW lines), and one each from the **Canadian Government Railway** and the **Atlantic Quebec & Western Railway**.

In 1919, the road number allotment retained **1016=1165** for the CaNoR 4-6-0s, and used **1166** for the lone CGR acquisition. In 1923, numbers **1167-1177** were assigned to GTR Ten Wheelers. In 1929, the number **1178** was assigned to a 4-6-0 from the AQ&W. All dates are mm-dd-yr.

| | 1015-1016 | | | | | 4-6-0 TEN WHEEL TYPE | | | | | | (G-1-A) |

Road numbers were not used in 1919 in order to keep the rest of the CaNoR 1000-series numbering and class sequence intact.
CaNoR 4-6-0s 1015-1016 are listed below as a reference.

			Specifications				Appliances		Weights	Fuel Capacity		Length	Notes
Cylinder	Gear	Driv.	Pressure	Boiler	T.E.	Haulage	Steam	Stkr.	Drivers/Eng./Total	Water	Coal		
18x24"	S	52"	140#	EST	17795	18%	sat		77/ 96/157060	2700 gals	tons	- '	[1000-1001]
18x24"	S	52"	180#	EWT?	20200	20%	sat		77/ 96/157060	2700 gals	tons	- '	[1000-1001 in 1911]
18x22"	S	50"	140#	EST	18507	18%	sat		77/ 96/ 000	gals	tons	- '	[1002]
18x24"	S	49"	150#	EST	20200	20%	sat		77/ 96/ 000	gals	tons	- '	[1005?-1006]
18x24"	S	56"	140#	EST	17024	17%	sat		70/100/ 000	gals	tons	- '	[1015]
18x24"	S	58"	140#	EST	18424	20%	sat		70/100/ 000	3000 gals	tons	- '	[1016]
18x24"	S	56"	140#	EWT	17024	17%	sat		70/100/ 000	gals	tons	- '	[143]
18x24"	S	56"	140#	EWT	17024	17%	sat		70/100/167300	2500 gals	tons	- '	[1032 in 1907]

Pittsburgh Locomotive & Car Company   1884

Serial	Shipped	New as	1899	1903	1904?	To	Acquired	1-1912	Disposition
727	4-12-84	BCR&N 78	142	CRI&P 1217	1117	*CaNoR 144?*	10-15-06	CaNoR G-1-A 1015	Sc 11- -16 PK
728	4-18-84	BCR&N 79	143	CRI&P 1206	1106	CaNoR 141	9-06-06	CaNoR F-2-A 1000	Sc 6-27-14 WER
729	4-18-84	BCR&N 80	144	CRI&P 1218	1118	CaNoR 147	11-06-06	CaNoR G-1-A 1016	Sc 12-16-16 PK
743	9-09-84	BCR&N 81	145	CRI&P 1207	1107	*CaNoR 145?*	9-26-06	CaNoR F-2-B 1002	Sc 9-04-17 PK
745	9-18-84	BCR&N 83	147	CRI&P 1209	1109	CaNoR 149	4-02-07	CaNoR F-2-E 1005	Sc 6-29-11
746	10-01-84	BCR&N 84	148	CRI&P 1210	1110	CaNoR 168	12-18-06	CaNoR F-2-E 1006	Sc 6-20-11 GG
747	10-09-84	BCR&N 85	149	CRI&P 1211	1111	CaNoR 142	9-06-06	CaNoR F-2-A 1001	Sc 11- -16 PK

Brooks Locomotive Company   1884

Serial	Shipped	New as	1899	1903	1904?	To	Acquired	1-1912	Disposition	To
1003	3- -84	BCR&N 91	150	CRI&P 1212	1112	CaNoR 143	9-06-06	—	Sc 4-21-11 PK	
1004	3- -84	BCR&N 92*	151	CRI&P 1213	1113	*CaNoR 146?*	9-26-06	CaNoR G-5-A 1032	Sc 1-19-16	DCo
1007	3- -84	BCR&N 95	154	CRI&P 1216	1116	*CaNoR 148?*	11-06-06	CaNoR G-5-B 1033		**CNR 1033**

CaNoR 1015-1016, 1032 and **CNR 1033** were four of ten locomotives purchased second-hand by the **Canadian Northern Railway** of **Chicago Rock Island & Pacific Railroad Company** ancestry through the Chicago-based dealer **J. T. Gardiner**. Records indicate CaNoR placed eight orders in 1906 with Gardiner for twenty-three locomotives consisting of two 0-6-0s, one 2-6-0, one 4-4-0 and nineteen 4-6-0s. Of the ten 4-6-0s in question, eight were ordered on August 8th 1906 and two on November 7th 1906. Their arrival (acquired) dates on CaNoR property, likely Fort Rouge in Winnipeg, are given in the roster.

From this group, only **1033** was taken into CNR stock, but all the former CaNoR road numbers and F and G classes, even though not active in 1919, were skipped in order to keep the rest of the CNR 1000-series numbering and class sequences intact.

The seven from Pittsburgh, delivered between December 1883 and October 1884, were from ten built in three lots as **Burlington, Cedar Rapids & Northern Railroad** 76 to 85, as illustrated by BCR&N 76. BCR&N 76 and

77 (#699-700) arrived in December 1883, 78-80 (#727-729) were shipped in April 1884, and BCR&N 81-85 (#743-747) in September and October 1884. They were later renumbered to BCR&N 140-149. After amalgamation, they were randomly renumbered into a CRI&P 1204 to 1218 and 1104 to 1118 series. Two of the lots had 50-inch drivers, and one, BCR&N 78-80, 57-inch drivers. It is believed any major change in driver size (*cf.* CaNoR 1000) occurred while in CRI&P service, but the other minor differences stem from varying methods of calculation and the locomotive's physical condition. The increase in boiler pressure was made by CaNoR in April 1911.

In March 1884, Brooks built five Ten Wheelers as **Burlington, Cedar Rapids & Northern Railroad** 91-95 (#1003-1007) which later were renumbered BCR&N 150-154. At amalgamation with the CRI&P, they were renumbered 1212-1216 and then 1112-1116.

If CaNoR records have been interpreted correctly, the ten from Rock Island were assigned CaNoR road

(text continues on next page)

numbers 141 to 149 in sequence following the broken CRI&P numbers, with the exception of 144 and 168. At this point, however, conflicting data exists between the CaNoR stock book record, locomotive diagram sheets and photographic evidence.

The railway's 1912 stock book listed the Brooks serial numbers for BCR&N 91, 92 and 95 as #1078, 1079 and 1082, but these are for BCR&N 86, 87 and 90 (later 160, 161 and 164). Confirmation of the serials used in the roster above was by magnifying the builder's plate in the photograph of BCR&N 92 (below) to ascertain it was indeed #1004. It may be that J. T. Gardiner and/or CaNoR personnel used the BCR&N 160-series Brooks builder's numbers (#1078-1082) to erroneously trace the heritage of these three second-hand 4-6-0s.

Another problem was with the CaNoR stock book entry showing CRI&P 1117 as the predecessor of both CaNoR 146 and 147, with CRI&P 1118 not listed. For many years evidence to clarify this duplication suggested CaNoR 146 was 1117, and 147 was 1118. Subsequently, in the CaNoR 1912 renumbering list, they were shown as G-5-A 1032 and G-5-B 1033 respectively.

Although not included in the group of seven locomotives sold to CaNoR, the first of the order, BCR&N 76, was chosen for the builder's photograph at Pittsburgh in December 1883. The order was shipped with several adornments favoured in the 1880s, notably the capped stack, polished boiler jacket and cylinder chest covers, and tender striping. None of these Ten Wheelers with evenly-spaced drivers would be taken into CNR stock.
[PITTSBURGH WORKS PHOTO P-63/ALCO HISTORIC PHOTOS]

However more recent investigation has shown this is not the case. In the stock book entry, the builder was listed as Pittsburg, but CaNoR diagram sheets for 1912 and 1919 show 143, 146/1032 and 148/1033 as Brooks

(text continues on next page)

Destined to spend a decade at Canadian Northern Railway construction sites, **BCR&N 92** was new at Dunkirk in the winter of 1884. It was fitted with more ornamentation than BCR&N 76 (above), especially the splash covers over the pilot truck wheels and striping on wheels, domes, headlight, cylinder chests and tender. Other niceties, once standard for power in the 1880s, would eventually be removed or replaced, in favour of more practical accessories. These victims of progress would include the wooden "cow-catcher" pilot with its pin-and-link coupling, and the large housing for the kerosene headlight.
[BROOKS WORKS PHOTO B-191/ALCO HISTORIC PHOTOS]

locomotives. This discrepancy was resolved when photographs of BCR&N 92 and CaNoR 1033 were compared. It became obvious both were Brooks-built 4-6-0s, primarily because of the unevenly-spaced drivers. It was not from Pittsburgh (as in the stock book), for those built there had evenly-spaced driving wheels as shown in the photo of BCR&N 76 (page G-3). As a result of this change in identities, the two former Brooks "orphans" were, by elimination, Pittsburghs, and tentatively assigned CaNoR road numbers based upon the differences in driving wheel diameters and the 1912 CaNoR classification. Unless additional information surfaces, the heritage of these 4-6-0s, especially CNR 1033, will remain uncertain.

CaNoR G-5-A 1032 was sold in January 1916 to the Dominion Collieries at Bienfait, Saskatchewan. CaNoR 1000 was sold to the Winnipeg Electric Railway Company, although the CaNoR stock book shows it sold to the Winnipeg Power Co. on July 7th 1914. The final dispositions of both former CaNoR 1000 and 1032 are unknown.

Class not issued by CNR   **G-1-a**

CNR 1016 (first)   4-6-0 TEN WHEEL TYPE   G-1-b

Cylinder	Gear	Driv.	Pressure	Boiler	T.E.	Haulage	Steam	Stkr.	Drivers/Eng./Total	Water	Coal	Length	Notes
19x24"	S	55"	140#	BEL	18760	19%	sat		100/130/216000	4000 gals	8 tons	54-11'	

Specifications / Appliances / Weights / Fuel Capacity headers span above.

Builder and build-date are unknown   (1) Acquired by CNR 9-01-1919

	Serial	Shipped	New as	by 1912	10-19-1912	2-1916?	Disposition
			—	—	—	G-1-B	
1016/1	???	-??	K&M ?	K&M 607	(CaNoR)–AS 607	CaNoR 1013/2	Sc 10- -23 MN

CNR 1016 (first) may have originally been a **Kanawha & Michigan Railroad** 4-6-0 which was acquired secondhand by the **Canadian Northern Railway**. Although CaNoR apparently purchased it in 1912, it was assigned and perhaps lettered as **Angus Sinclair** 607 until 1916. The CaNoR stock book also lists it as being built by the Rome Locomotive Works in 1903, but the entry is spurious as that builder was forced to close in 1893 as the result of a fire. The *CNR Mechanical Department*

Locomotive Diagrams record both the build- and rebuild-dates as 1903. Further investigation has ascertained the sheet contains mixed-up histories for CaNoR first and second 1013. (See CaNoR 1013 and 1014 in class F-5-A on page F-8). CaNoR second 1013 was the 4-6-0 rebuilt by the East St.Louis Locomotive Works in 1911 – as recorded under note "GA" for CaNoR second 1013 on page 77 of *Canadian National Steam Power* (1969).

CNR 1016 (second)   4-6-0 TEN WHEEL TYPE

See Class F-1-c (page F-8).

CNR 1017-1020 (first)   4-6-0 TEN WHEEL TYPE   G-2-a

Cylinder	Gear	Driv.	Pressure	Boiler	T.E.	Haulage	Steam	Stkr.	Drivers/Eng./Total	Water	Coal	Length	Notes
18x24"	S	54"	150#		00		sat		85/108/ 000	gals	tons	- '	[ATSF]
18x24"	S	56"	180#		21245	21%	sat		91/118/197900	3460 gals	tons	- '	[CaNoR 1909]

Rhode Island Locomotive Company   1890   (4) Acquired by CNR 9-01-1919

	Serial	Shipped	Ordered as	New as	1898	1900	8-01-1902	1-1912	Disposition
				—	—	—	—	G-2-A	
1017/1	2392	6- -90		ATSF 463	ATSF 503	ATSF 145	CaNoR 63	CaNoR 1017	Sc 4-29-20 PK
1018/1	2411	6- -90		ATSF 467	ATSF 507	ATSF 149	CaNoR 64	CaNoR 1018	Sc 6-15-21 PK
1019	2390	6- -90	SCRR —	ATSF 461	ATSF 501	ATSF 143	CaNoR 65	CaNoR 1019	Sc 4-24-20 PK
1020	2393 (2388)	6- -90		ATSF 464	ATSF 504	ATSF 146	CaNoR 66	CaNoR 1020	Sc 6-30-20 PK

CNR 1017-1020 were part of an order for eight locomotives built as **Atchison Topeka & Santa Fe Railroad** 461-464 (#2390-2393); 465 and 466 (#2388 and 2389); and 467 and 468 (#2411 and 2412). Even though ATSF 461 (**CNR 1019**) was ordered and built for the **South Carolina Railroad**, it became part of the ATSF order. Rhode Island records show ATSF 464 was built under serial 2393, but CaNoR records show it as 2388 (ie. ATSF 465/147). Rather than suspecting an error in documentation, railway historians believe the boilers were switched some time before ATSF 146 became a **Canadian Northern Railway** locomotive. **CaNoR 1018** pulled the inaugural train from Victoria on September 9th 1918 to open the Canadian Northern Pacific line on Vancouver Island.

CaNoR 64 (1018) at an unidentified location, perhaps in the fall of 1902 (based upon its rather pristine appearance), would spend a great deal of the next twenty years in Canada working the construction camps. [JOHN A. REHOR/GEORGE CARPENTER COLLECTION]

See Figure FG (page F-2) for a summary of the multiple use of 4-6-0 road numbers 1000-1018 between 1912 and 1950 (page F-2).

These former Santa Fe Ten Wheelers, with their narrow fireboxes set between the second and third drivers, were never superheated. CaNoR 1017, likely in the Transcona scrap line about 1920, attested to the demanding working conditions and few technological upgrades which would relegate these nineteenth century 4-6-0s to early retirement. The retrofitted outside steam pipes were uncommon on saturated engines. [H.L. GOLDSMITH/GEORGE CARPENTER COLLECTION]

CNR 1021-1024   4-6-0 TEN WHEEL TYPE   G-2-b

Cylinder	Gear	Driv.	Specifications Pressure	Boiler	T.E.	Haulage	Appliances Steam	Stkr.	Weights Drivers/Eng./Total		Fuel Capacity Water	Coal	Length	Notes
18x24"	S	55"	160#		19200		sat		81/108/	000	gals	tons	- '	[TSL&KC 27=49]
18x24"	S	54"	160#		19600		sat		81/108/	000	gals	tons	- '	[TSL&KC 54, 69]
18x24"	S	56"	160#		18880		sat		81/108/	000	gals	tons	- '	[TSL&W 62=82]
18x24"	S	56"	150#		18360	18%	sat		81/108/	000	gals	tons	- '	[CaNoR]

Rhode Island Locomotive Company   1888-1892   (3) Acquired by CNR 9-01-1919

	Serial	Shipped	In service	New as		1901	5-11-1907	1-1912		Disposition
				—	—	E (1904)		G-2-B	F-4-A	
1021	2142	2-89	3-07-89		**TSL&KC 41**	**TSL&W 76**	(JTG) CaNoR 156	**CaNoR 1021**		Sc -20 EH
—	2196	2-90	3-04-90		TSL&KC 47	TSL&W 82	(JTG) CaNoR 94		CaNoR 1011	To CNR (1014)
—	2543	6-91	6-15-91		TSL&KC 54	TSL&W 54*	(JTG) CaNoR 93		CaNoR 1010	To **CNR 1013**/1
1022	2053	10-88	10-23-88	MCCo 22	TSL&KC 27	TSL&W 62	(JTG) CaNoR 157	**CaNoR 1022**		Sc 7-26-20 PK
1023	2718	9-92	9-28-92		TSL&KC 69	TSL&W 69*	(JTG) CaNoR 158	**CaNoR 1023**		Sc -20 AK
1024	2054	10-88	10-23-88	MCCo 23	TSL&KC 28	TSL&W 63	(JTG) CaNoR 159	**CaNoR 1024**		Sc 7-11-17

*:TSL&W 54 and 59 classed Ea in 1904.

CNR 1021-1024 and F-4-a **1013** and (**1014**) were six 4-6-0s purchased through the dealer **James T. Gardiner** from the **Toledo, St. Louis & Western Railroad** in 1907. They had been built for the TStL&W's predecessor, the **Toledo, St. Louis & Kansas City Railroad**. TStL&KC 40-44 were built under serials #2141-2145, 47-49 under #2196-2198, 51-54 under #2540-2543, with 69 and 70 under #2718 and 2719. TStL&KC 27-29 were built as **Macon Construction 22-24**, but if the TStL&KC in-service date is correct, they may have been diverted from the builder to the TStL&KC as 27-29. The dealer, CaNoR or historians apparently confused road numbers 62 and 63 from the "Kansas City" and "Western" rosters. Some publications, including Clegg and Corley: *Canadian National Steam Power* (p. 77) record the build-data for TStL&KC 62-65 as Rhode Island #2824-2827, built in August 1892. They were larger and heavier 4-6-0s than those purchased by the CaNoR. The pair became TStL&W E-1 100 and 101, which were documented as removed from the TStL&W roster in 1922 and 1920.

CaNoR 159 (1024) was photographed at an unidentified construction site. Despite the fogged right-hand side of the photograph, the freshly painted tender, clean CaNoR lettering and number suggest a date of about May 1907, or at least soon after its purchase from the secondhand dealer. Narrow fireboxes and ash pans set in between the driving wheel frames were characteristic of nineteenth-century technology. Raking ashes from the shallow pans was a constant task for both hostlers and crews. Access to remove ash was between the unevenly-spaced rear drivers. Bought second-hand for construction of the line, these saturated, wooden cab CaNoR G-2 4-6-0s were the first to be removed from service.
[JOHN A. HINDS/IAN WEBB COLLECTION]

TStL&W 62, 63 and 76 (**1022**, **1024** and **1021**) were recorded retired in 1905 before being sold to Gardiner. TStL&W 82 (**1014**) was shown in the company record as sold to the **Cincinnati, Bluffton & Chicago Railroad**, likely in 1905, before the sale to Gardiner and the **Canadian Northern Railway** in 1907. In 1920, both CNR 1013 and CaNoR 1011 were towed to Moncton Reclamation yard for dismantling.

1025						4-6-0 TEN WHEEL TYPE								G-2-B
			Specifications				Appliances		Weights		Fuel Capacity		Length	Notes
Cylinder	Gear	Driv.	Pressure	Boiler	T.E.	Haulage	Steam	Stkr.	Drivers/Eng./Total		Water	Coal		
18x24"	S	55"	150#	WT	18000	18%	sat		82/100/178000		3000 gals	tons	- '	[CaNoR 1913]

Rhode Island Locomotive Company		1891					
	Serial	Shipped	New as	7-19-1906	1-1912		Disposition
			—	—	G-2-B		
	2547	6- -91	LLR 2/2	CNQ 51	CaNoR 1025		Sc 6-14-17 GG

CaNoR 1025 would have become CNR G-2-b 1025 but was scrapped at Parry Sound (South Parry) before the formation of the CNR. It had been built for the **Lower** **Laurentian (Bas Laurentides) Railway** and became the property of the **Canadian Northern Quebec Railway** at the time of its formation in 1906.

CNR 1026						4-6-0 TEN WHEEL TYPE								G-2-c
			Specifications				Appliances		Weights		Fuel Capacity		Length	Notes
Cylinder	Gear	Driv.	Pressure	Boiler	T.E.	Haulage	Steam	Stkr.	Drivers/Eng./Total		Water	Coal		
18x24"	S	54"	150#	WT	20486	20%	sat		82/110/178000		3400 gals	tons	- '	[CaNoR 1909]

Rhode Island Locomotive Company		(unknown)				(1) Acquired by CNR 9-01-1919	
	Serial	Shipped	New as	10-01-1906	1-1912		Disposition
			—	—	G-2-C		
1026	????	-90?	Not known	(JTG) CaNoR 166	CaNoR 1026		Sc 7-08-20 PK

The origins of **1026** are not given in company records. It came to the **Canadian Northern Railway** via second-hand dealer **James T. Gardiner** of Chicago (Heights), Illinois. One CaNoR record gives the driver diameter as 49 inches, three inches smaller than the minimum size assigned G class 4-6-0s.

CNR 1027-1030						4-6-0 TEN WHEEL TYPE								G-3-a
			Specifications				Appliances		Weights		Fuel Capacity		Length	Notes
Cylinder	Gear	Driv.	Pressure	Boiler	T.E.	Haulage	Steam	Stkr.	Drivers/Eng./Total		Water	Coal		
18x24"	S	57"	190#	WT	22032	22%	sat		99/120/220000		4500 gals	6 tons	56-9'	[CaNoR 1913]
18x24"	S	57"	190#	WT	20400	21%	sat		99/120/220000		4500 gals	6 tons	56-9'	[CNR]

Canadian Locomotive Company		1907	$16,494			(4) Acquired by CNR 9-01-1919	
	Serial	Shipped	New as	9-1909	1-1912		Disposition
			—	—	G-3-A		
1027	785	12-05-07	COR 17*	MM&Co (CaNoR)–COR 17	CaNoR 1027		Sc 5-30-25 MV
1028	786	12-30-07	COR 18	MM&Co (CaNoR)–COR 18	CaNoR 1028		Sc 9-28-34 HQ
1029	787	12-31-07	COR 19	MM&Co (CaNoR)–COR 19	CaNoR 1029		Sc 6-02-33 HQ
1030	788	1-06-08	COR 20	MM&Co (CaNoR)–COR 10	CaNoR 1030		Sc 5-30-25 MV

CNR 1027-1030 were built for the **Central Ontario Railway**. The January 1907 order was originally for two locomotives, but within a month was increased to four. They were acquired in the purchase of the COR by **Mackenzie, Mann & Company Limited**, renumbered into the **Canadian Northern Railway** system in 1912 and, in July 1914, transferred to CaNoR ownership.

COR 17 (1027), at Kingston in December 1907, was the first of four medium-sized 4-6-0s built for the eastern Ontario railway. Typical of the smaller Ten Wheelers of the time designed for slower speeds and rougher track conditions, they retained their small fireboxes between the rear drivers and remained saturated. Two were retired after only nineteen years of service, while the other two remained in service until the early 1930s.
[CLC, MILN-BINGHAM LITHOGRAPH FROM A HENDERSON PHOTOGRAPH/ DON McQUEEN COLLECTION]

G-2-B
G-2-c
G-3-a

CNR 1031 — 4-6-0 TEN WHEEL TYPE — G-4-a

Cylinder	Gear	Driv.	Pressure	Boiler	T.E.	Haulage	Steam	Stkr.	Drivers/Eng./Total	Water	Coal	Length	Notes
18x24"	S	54"	180#	WT	21631	22%	sat		88/120/200000	3320 gals	6 tons	- '	[CaNoR 1909]
18x24"	S	54"	180#	WT	21631	22%	sat		88/120/200000	4000 gals	5 tons	- '	[CaNoR 1913]

Baldwin Locomotive Works – Burnham, Williams & Company			1907					(1) Acquired by CNR 9-01-1919
	Serial	Shipped	Delivered	Ordered as	New as	1-1912		Disposition
				—	—	G-4-A		
1031	30557	4- -07	5-26-07	M&TR – (JTG)	CaNoR 140	CaNoR 1031		Sc 12-15-20 PK

CNR 1031 was bought by the **Canadian Northern Railway** for $13,500 through the agency of **James T. Gardiner** of Chicago, Illinois. It had been originally ordered by a proposed Mexican railway, the **Morelia & Tacambara Railway**, but never delivered when the scheme collapsed. CaNoR records show it arriving on May 26th 1907.

CNR 1032 — 4-6-0 TEN WHEEL TYPE — G-5-A

See note under (CaNoR) **1015-1016** (page G-2).

Class not issued by CNR — G-5-a

CNR 1033 — 4-6-0 TEN WHEEL TYPE — G-5-b

Cylinder	Gear	Driv.	Pressure	Boiler	T.E.	Haulage	Steam	Stkr.	Drivers/Eng./Total	Water	Coal	Length	Notes
18x24"	S	57"	140#	EWT	16234	16%	sat		/ / 000	3000 gals	tons	- '	[CaNoR 1907]
18x24"	S	57"	140#	EFT	16234	16%	sat		/ / 000	3000 gals	tons	- '	[CaNoR 1920]

Brooks Locomotive Company		1884					(1) Acquired by CNR 9-01-1919
	Serial	Shipped	New as	1899	1903	1904?	9-06-1906	1-1912	Disposition
			—	—	—	—	—	G-5-B	
1033	1007	3- -84	BCR&N 95	BCR&N 154	CRI&P 1216	CRI&P 1116	CaNoR 148?	CaNoR 1033	Sc 7-21-20 PK

CNR 1033. See explanatory notes and photograph under (CaNoR) **1015-1016** (page G-2).

CaNoR 1033, the last remaining second-hand steamer from the BCR&N, was photographed near the end of its career about 1918 or 1919. If so, it was at Fort Rouge before relettering into CNR livery – if that is indeed the purpose of the painted (or chalked) "1033" on the side of the uncoupled tender. Another interpretation of the photograph, based on the removal of most appliances as well as the cross-head and guides, suggests the 4-6-0 was in the Fort Rouge scrap line about 1920. The unequally-spaced drivers and boiler features provide the evidence necessary to confirm the 4-6-0 was a Brooks product, rather than one from Pittsburgh as the CaNoR record would have it.
[CANOR PHOTO/GEORGE CARPENTER COLLECTION]

CNR 1034-1038   4-6-0 TEN WHEEL TYPE   G-6-a

Cylinder	Gear	Driv.	Pressure	Boiler	T.E.	Haulage	Steam	Stkr.	Drivers/Eng./Total	Water	Coal	Length	Notes
			Specifications				Appliances		Weights			Length	Notes
19x24"	S	57"	180#	WT	23256	23%	sat		111/141/261000	5000 gals	10 tons	59-4'	[CaNoR 153-154; 1907]
19x24"	S	57"	180#	WT	23260	23%	H-C		111/141/261000	5000 gals	11 tons	59-4'	[CNR]

	Serial	Delivered	New as	11-1905	1-1912	Superheated	Disposition
Canada Foundry Company Ltd. – Davenport Works		1905	$15,000				(5) Acquired by CNR 9-01-1919
			—		G-6-A		
1034	836	10-06-05	JBR 151		CaNoR 1034	6-17 PK	Sc 8-31-25 PK
1035	837	10-24-05	JBR 152		CaNoR 1035	5-17 PK	Sc 8-11-25 PK
1036	838	11-04-05	JBR 153*	CaNoR 153	CaNoR 1036	8-16 PK	Sc 3-26-26 PU
1037	839	11-08-05	JBR 154	CaNoR 154	CaNoR 1037	6-16 PK	Sc 8-20-25 PK
1038	840	11-26-05	JBR 155		CaNoR 1038	1-17 PK	Sc 3-15-26 PU

CNR 1034-1038, built for the **James Bay Railway**, were ordered on January 9th 1905 by the Canadian Northern. Very soon after delivery, two from the order were transferred to CaNoR and, by 1907, were known to have been working in the Prince Albert district of Saskatchewan. Although the *CNR Mechanical Department Locomotive Diagrams* show the builder's serials as #836-840, Canada Foundry records also have CPR 2-8-0C M2F 1502-1505 (3302-3305 after 1912), built in 1904-1905, assigned #836-839. Both **CNR 1036** and **1038** were planned for removal from the roster in 1925, but were returned to service later that year.

JBR 151 (1034), in the aggregate pit at Falding, Ontario (five miles south of South Parry), likely in the summer of 1906, left the Toronto builder saturated, fitted with one-piece inclined cylinder castings, piston valves and inside steam pipes. Unlike the pair transferred to the CaNoR and worked west, the remaining three kept their James Bay Ry lettering until the CaNoR classification and renumbering scheme went into effect in 1912.
[ANDREW MERRILEES COLLECTION/
LIBRARY AND ARCHIVES CANADA PAC 165555]

4-6-0 TEN WHEEL TYPE												G-7-A
Specifications						Appliances		Weights	Fuel Capacity		Length	Notes
Cylinder	Gear	Driv.	Pressure	Boiler	T.E.	Haulage	Steam	Stkr.	Drivers/Eng./Total	Water	Coal	Length	Notes
18x24"	S	56"	#		00		sat		/ / 000	gals	tons	- '	[orig]
19x24"	S	56"	150#		17704	17%	sat		86/115/185000	2000 gals	8 tons	59-9'	[164 CaNoR, 1907]
19x24"	S	57"	150#		19380	19%	sat		89/115/185000	2500 gals	8 tons	59-9'	[165 CaNoR, 1907]
19x24"	S	50"	140#		18000	18%	sat		89/115/185000	2000 gals	8 tons	59-9'	[1039-1041 CaNoR, 1912]

Brooks Locomotive Company	1887	(264)						
	Serial	Shipped	New as	Rblt 2-13-1891	9-13-1905	8-23-1906	1-1912	Disposition
			—		—	—	G-5-A	
	1262	8- -87	BR&P 40	Rome Loco. Works		(AECO) CaNoR 164	CaNoR 1039	Sc 8-06-17 PK
	1263	8- -87	BR&P 41			(AECO) CaNoR 165	CaNoR 1040	Sc 4-12-17 PK
	1259	8- -87	BR&P 37	Rome Loco. Works	JBR 101		CaNoR 1041	Sc 6-14-17 GV

Ten Wheelers **1039-1041** would have become **CNR** G-7-a **1039-1041** had they not been scrapped in 1917. They had been part of an order built for the Buffalo Rochester & Pittsburgh Railroad as BR&P 37-43 (#1259-1265). Canadian Northern Railway purchased three from the order in two lots, one in 1905 for the James Bay Railway, and the other two in 1906 for its western lines. The latter two were purchased through the Atlantic Equipment Company, a dealer in New York City, although the locomotives were shipped from Du Bois, Pennsylvania.

BR&P 38 was one of four in a Brooks order destined to remain south of the Canadian border. Of the seven built, the three which crossed the border to work for the CaNoR were removed from the roster before the formation of the CNR. Posed for the photographer at Dunkirk, New York in August 1887, BR&P 38 provided evidence as to the appearance of the order, although two of the three CaNoR purchased may have been physically modified during the rebuild at Rome, New York, but, as built, the 4-6-0 represented the mainstream in late nineteenth century technology with a capped stack, extended smokebox, wagon top boiler, unevenly-spaced drivers providing ash pan access, and moderate pinstriping.
[BROOKS WORKS PHOTO B-264/ALCO HISTORIC PHOTOS]

CNR 1042					4-6-0 TEN WHEEL TYPE								G-8-a
Specifications						Appliances		Weights	Fuel Capacity		Length	Notes	
Cylinder	Gear	Driv.	Pressure	Boiler	T.E.	Haulage	Steam	Stkr.	Drivers/Eng./Total	Water	Coal	Length	Notes
20x26"	S	56"	200#	WT	31600	32%	sat		113/140/220000	4000 gals	8 tons	59-3'	[CaNoR 1913]
19x26"	S	57"	180#	WT	25193	26%	H-C		113/140/220000	4000 gals	10 tons	59-3'	[CNR 1924]

Baldwin Locomotive Works – Burnham, Williams & Company	1901					(1) Acquired by CNR 9-01-1919	
	Serial	Shipped	New as	11-1906	1-1912	Superheated	Disposition
			—	—	G-8-A	EsC	
1042	18541	1- -01	QLSJ 17	MM&Co (CaNoR)–QLSJ 17	CaNoR 1042	8-18 PK	Sc 11-08-33 LM

CNR 1042 was built for the **Quebec & Lake St. John Railway**, before being acquired with the purchase of the QLSJ by **Mackenzie, Mann & Company Limited**, re-numbered into the **Canadian Northern Railway** system in 1912 and, in July 1914, transferred to CaNoR ownership. CaNoR 1042 was equipped with Economy (Universal) steam chests (EsC) when it was superheated.

CNR 1042, likely at London in 1933 on its way to the scrap yard, was built for the Q&LSJ at least thirty years before. Although delivered with the narrow firebox and unevenly-spaced drivers in vogue at the turn of the century, one of the elements contributing to its longevity was it had been superheated. It and G-14-a 1109 were the oldest CaNoR 4-6-0s to be superheated, albeit with Economy steam chests.
[LAWRENCE A. STUCKEY/WES DENGATE COLLECTION]

CNR 1043-1047 — 4-6-0 TEN WHEEL TYPE — G-9-a

Cylinder	Gear	Driv.	Pressure	Boiler	T.E.	Haulage	Steam	Stkr.	Drivers/Eng./Total	Water	Coal	Length	Notes
			Specifications				Appliances		Weights	Fuel Capacity			
19x24"	S	56"	180#	BEL	23670	24%	sat		107/133/208000	4000 gals	8 tons	59-5½'	[CaNoR 1909]
19x24"	S	57"	180#	BEL	23256	23%	sat		107/137/220000	4000 gals	8 tons	59-5½'	[af. 1924]

Baldwin Locomotive Works – Burnham, Williams & Company   1901 $11,350   (5) Acquired by CNR 9-01-1919

	Serial	Shipped	New as	9-1904	1-1912	11-1925	Disposition	To
			—	—	G-9-A			
1043	19506	9- -01	CaNoR 36*	CaNoR 100/2	CaNoR 1043		Sc 10-14-25 PK	
1044	19507	9- -01	CaNoR 37	CaNoR 101/2	CaNoR 1044		Sc 10-30-25 PK	
1045	19624	10- -01	CaNoR 38	CaNoR 102	CaNoR 1045		Sc 9-30-25 SH	
1046	19625	10- -01	CaNoR 39	CaNoR 103	CaNoR 1046	(CaNoR 103)	Dn11-30-25 W	**Edm**
1047	19626	10- -01	CaNoR 40	CaNoR 104	CaNoR 1047		So12- -25 W	**CJCo**

CNR 1043-1047 were ordered by the **Canadian Northern Railway** on March 6th 1901. **CNR 1046** was donated to the **City of Edmonton** on November 30th 1925 because two decades earlier it had been the first locomotive to enter the city using the CaNoR main line, on November 24th 1905. As CaNoR 103, it was slated for display, but remained stored at the exhibition grounds until it was returned to CNR in either 1937 or 1938 (no date is recorded) and put into records for scrapping on April 20th

CaNoR 36 (1043), at Philadelphia in September 1901, was one of five in the first order for new 4-6-0s placed by the Canadian Northern Railway. The design included elongated Belpaire fireboxes which extended beyond the rear pair of drivers. Despite these design improvements, none of the group was ever superheated.
[BLW PHOTO 1510/H.L. BROADBELT/WES DENGATE COLLECTION]

1938 at the Edmonton (Calder) roundhouse. **CNR 1047** was sold for $800 to the **Canadian Junk Company**, of Vancouver, British Columbia.

G-8-a

G-9-a

CNR 1048-1052   4-6-0 TEN WHEEL TYPE   G-10-a

			Specifications				Appliances		Weights	Fuel Capacity		Length	Notes
Cylinder	Gear	Driv.	Pressure	Boiler	T.E.	Haulage	Steam	Stkr.	Drivers/Eng./Total	Water	Coal		
19x24"	S	56"	180#	BEL	23670	24%	sat		121/139/237700	4000 gals	8 tons	58-4'	[CaNoR]
19x24"	S	56"	180#	BEL	23256	23%	H-C		121/139/238000	4000 gals	8 tons	58-4'	[af. 1924]
19x24"	S	57"	180#	BEL	23256	23%	H-C		121/139/238000	3500 gals	8½ tons	58-4'	[1050]

Canadian Locomotive Company		1902	$14,890					(5) Acquired by CNR 9-01-1919	
	Serial	Shipped	New as	9-1904	1-1912	Superheated	Stl cab	Disposition	To
			—	—	G-10-A				
1048	542	2-05-02	CaNoR 31	CaNoR 105	CaNoR 1048	1-21 MV		So 1- -28 HJ	A&J
1049	543	2-05-02	CaNoR 32	CaNoR 106	CaNoR 1049	7-20 PK #		Sc 5-07-37 AK	
1050	544	2-05-02	CaNoR 33	CaNoR 107	CaNoR 1050	8-20 PK #		Sc 7-16-36 PU	
1051	545	2-11-02	CaNoR 34	CaNoR 108	CaNoR 1051			Rs 3- -26 So 8-16-26 PK	MP&P
1052	546	2-11-02	CaNoR 35*	CaNoR 109	CaNoR 1052	12-20 PK #	6-32 AV	Sc 6-12-45 AK	

#: offset exhaust (steam) pipe.

CNR 1048-1052 were ordered by the **Canadian Northern Railway** on May 30th 1901. In 1928, CNR 1048 was sold for $10,600 (from Lindsay) to the **Alma & Jonquieres Railway**, likely as its 1048, and was scrapped in 1931.

CNR 1051 was sold on August 16th 1926 for $3000 plus repair costs, to the **Manitoba Power & Paper Company** at Pine Falls, but its final disposition is unknown.

CNR 1053-1057   4-6-0 TEN WHEEL TYPE   G-10-a

			Specifications				Appliances		Weights	Fuel Capacity		Length	Notes
Cylinder	Gear	Driv.	Pressure	Boiler	T.E.	Haulage	Steam	Stkr.	Drivers/Eng./Total	Water	Coal		
19x24"	S	56"	180#	BEL	23670	23%	sat		121/139/237700	4000 gals	8 tons	59-1'	[CaNoR 1909]
19x24"	S	57"	180#	BEL	23256	23%	H-C		121/139/238000	4000 gals	8 tons	58-4'	[af. 1924]
19x24"	S	57"	180#	BEL	23256	23%	H-C		121/139/242000	3500 gals	7 tons	58-4'	[1053]

Canadian Locomotive Company		1902	$14,900						(5) Acquired by CNR 9-01-1919	
	Serial	Shipped	New as	9-1904	1-1912	Superheated	Stl cab	Tender change	Disposition	To
			—	—	G-10-A					
1053	561	8-18-02	CaNoR 51	CaNoR 110	CaNoR 1053	11-22 PK		bf-28	So 6- -28 RG	NAC
1054	562	9-26-02	CaNoR 52	CaNoR 111	CaNoR 1054				Sc 12-16-35 AK	
1055	563	9-01-02	CaNoR 53*	CaNoR 112	CaNoR 1055	8-22 PK	1-37 AK		Sc 9-07-51 AK	
1056	564	9-12-02	CaNoR 54	CaNoR 113	CaNoR 1056				So 11- -26 PK	CHA
1057	565	9-29-02	CaNoR 55	CaNoR 114	CaNoR 1057				Sc 5-03-27 PU	

Likely in the Moncton scrap line on November 11th 1935, **1054** was one of several in the class never superheated. This was one of the factors which eventually led to their retirement before World War II. [AL PATERSON & DON McQUEEN COLLECTIONS]

CNR 1053-1057 were ordered by the **Canadian Northern Railway** in January 1902. By the late 1920s, CNR 1053 had received a tender with a smaller capacity, and by the

late 1930s, CNR 1055 had received a steel cab. In 1928, CNR 1053 was sold for $3500 to the **North American**

(text continues on next page)

Collieries of Nacmine, Alberta, and later was sold to **Ajax Coal Company** at Medicine Hat before being scrapped. In 1926, **CNR 1056** was sold to the **Carter-Halls-Aldinger Company** of Winnipeg, and was delivered to a work site at Paddington, Manitoba before being scrapped, likely by CNR at Transcona in 1933. The $3000 purchase price was to be paid on a rental basis at the rate of $17.25 per day

The last surviving member of the class, superheated **1055** at Truro in October 1946, had been rebuilt with a steel cab, a change which undoubtedly contributed to the extension of its service life to 1951. [AL PATERSON COLLECTION]

until the total was accrued. The disposition of 1054's tender, not scrapped with the engine at Moncton in 1936, is not known.

CNR 1058-1082 — 4-6-0 TEN WHEEL TYPE — G-10-b

Cylinder	Gear	Driv.	Pressure	Boiler	T.E.	Haulage	Steam	Stkr.	Drivers/Eng./Total	Water	Coal	Length	Notes
			Specifications				Appliances		Weights	Fuel Capacity			
19x24"	S	56"	180#	BEL	23670		sat		112/138/244000	4000 gals	9 tons	59-1'	[CaNoR 1909]
19x24"	S	56"	180#	BEL	23670	24%	sat		107/133/179000	4000 gals	9 tons	59-1'	[CaNoR 1913]
19x24"	S	57"	180#	BEL	23256	23%	H-C		109/135/237600	4500 gals	9 tons	58-4'	[af. 1924]
19x24"	S	57"	180#	BEL	23256	23%	H-C		109/135/237600	4500 gals	8½ tons	58-4'	[with hopper]

Canadian Locomotive Company 1903 $16,097 (25) Acquired by CNR 9-01-1919

	Serial	Shipped	Delivered	New as	1-1912	Superheated	Stl cab	Sold to	Disposition	To
				—	G-10-A			Tender to		
1058	586	6-30-03	7-08-03	CaNoR 115	CaNoR 1058				Sc 5-03-27 PU	
1059	587	7-06-03	7-13-03	CaNoR 116	CaNoR 1059	8-22 PK			Sc 5-31-37 PU	
1060	588	7-11-03	7-20-03	CaNoR 117	CaNoR 1060				So 2- -27 PK	WDC
1061	589	7-17-03	7-22-03	CaNoR 118	CaNoR 1061	9-23 EH	11-25 HQ		Sc 6-10-31 HQ	
1062	590	7-24-03	7-31-03	CaNoR 119	CaNoR 1062	11-22 EH	4-30		Sc 12-31-35 HW	
1063	591	8-01-03	8-08-03	CaNoR 120	CaNoR 1063				Sc 9-27-35 AK	
1064	592	8-06-03	8-13-03	CaNoR 121	CaNoR 1064	11-20 PK #	3-34 PK		Sc 9-30-40 AK	
1065	593	8-13-03	8-19-03	CaNoR 122	CaNoR 1065	3-21 MV		OCS	Sc 7-25-40 LM	
1066	594	8-22-03	8-29-03	CaNoR 123	CaNoR 1066	10-23 EH			So 2-03-26 C	A&J
1067	595	8-31-03	9-05-03	CaNoR 124	CaNoR 1067	3-22 PK			Sc 8-28-30 PU	
1068	596	9-07-03	9-12-03	CaNoR 125	CaNoR 1068			DP&P> CNR	Sc 6-17-29 EH	
1069	597	9-14-03	9-21-03	CaNoR 126	CaNoR 1069	8-21 MV	3-29		Sc 6-10-31 HQ	
1070	598	9-22-03	9-29-03	CaNoR 127	CaNoR 1070	8-23 EH			Sc 8-31-40 AK	
1071	599	9-26-03	10-01-03	CaNoR 128	CaNoR 1071	8-22 MV	10-37 HQ		Sc 12-20-47 JD	
1072	600	9-30-03	10-08-03	CaNoR 129	CaNoR 1072				Sc 8-02-27 MV	
1073	601	10-07-03	10-13-03	CaNoR 130	CaNoR 1073	10-23 MV			Sc 10-10-35 JD	
1074	602	10-16-03	10-21-03	CaNoR 131	CaNoR 1074	12-21 EH	12-29		Sc 10-31-35 HW	
1075	603	10-22-03	10-28-03	CaNoR 132	CaNoR 1075	12-20 EH #			Sc 4-30-37 AK	
1076	604	10-27-03	11-09-03	CaNoR 133	CaNoR 1076			OCS	Sc 12-16-35 AK	
1077	605	10-31-03	11-08-03	CaNoR 134	CaNoR 1077	10-20 PK #			Sc 8-31-40 AK	
1078	606	11-09-03	11-13-03	CaNoR 135	CaNoR 1078				Sc 9-01-27 MV	
1079	607	11-12-03	11-18-03	CaNoR 136	CaNoR 1079			OCS	Sc 10-31-35 AK	
1080	608	11-18-03	11-22-03	CaNoR 137	CaNoR 1080	10-21 EH	1-27 & 6-34		Sc 10-09-35 JD	
1081	609	11-24-03	11-27-03	CaNoR 138	CaNoR 1081	4-21 EH			Sc 3-16-37 MQ	
1082	610	11-30-03	12-04-03	CaNoR 139	CaNoR 1082				Sc 9-01-27 MV	

#: offset exhaust (steam) pipe.

G-10-a

G-10-b

CNR 1058-1082 were ordered by the Imperial Rolling Stock Company for the **Canadian Northern Railway** in 1903. Four other Ten Wheelers (CNR 1207-1210) were ordered at the same time. In 1925, **CNR 1066** was slated for retirement, but was returned to service the same year, and sold for $5000 to the **Alma & Jonquieres Railway**. A&J 1066 was scrapped in 1941. In 1927, **CNR 1060** was sold for $3000 to the **Western Dominion Coal Company**,

(text continues on next page)

One of the first large lots of new motive power acquired by the Canadian Northern at the turn of the century was this group of saturated 4-6-0s which ultimately became CNR's G-10-b class. Those not subsequently superheated, such as **1058**, likely at Transcona in 1927, met the scrapper's torch in the 1920s.
[CNR LOCOMOTIVE DATA CARD]

CNR 1074, at Turcot on May 13th 1928, had yet to undergo alterations to either the headlight and bell location; nor had it yet acquired a boiler tube pilot or a rebuilt slab-sided tender.
[H.L. GOLDSMITH/GEORGE CARPENTER COLLECTION]

One of the two locomotives in the class rebuilt with a steel cab was **1071**, photographed at Spadina about 1944.
[AL PATERSON COLLECTION]

Other alterations included a larger capacity water pump and a relocated air reservoir. Despite these appliance changes, all twenty-five in the class retained their Belpaire fireboxes and unevenly-spaced driving wheels.

G-10-b

but its disposition is unknown. That same year, on February 27th, **CNR 1068** was sold for $6000 and exchanged for CNR E-7-a 847 at the **Donnacona Pulp & Paper Company**, only to be returned on August 28th 1928 to CNR in trade for a more practical 0-6-0, O-8-a 7099. Before the 1927 sale, the CaNoR ownership of 1068 was

changed to CNR. CNR put the 4-6-0 back in service until its retirement in 1929.

The disposition of tenders for **1076** and **1079**, not scrapped with the engines at Moncton in 1935, is not known. However, the tender from **1065** was set aside as a water transport car after the engine was retired.

CNR 1083-1102 — 4-6-0 TEN WHEEL TYPE — G-11-a

Cylinder	Gear	Driv.	Specifications Pressure	Boiler	T.E.	Haulage	Appliances Steam	Stkr.	Weights Drivers/Eng./Total	Fuel Capacity Water	Coal	Length	Notes
19x26"	S	57"	200#	WT	27993	28%	sat		137/159/263300	5000 gals	10 tons	62-3½'	[CaNoR 1909-1913]
19x26"	S	57"	190#	WT	26600	26%	H-C		116/164/268400	5000 gals	10 tons	62-3½'	[CaNoR/CNR]

Canada Foundry Company Ltd. – Davenport Works   1906 $17,140   (20) Acquired by CNR 9-01-1919

	Serial	Delivered	New as	1-1912	Superheated	Disposition
			—	G-11-A		
1083	849	9-27-06	CaNoR 300	CaNoR 1083	3-17 PK	Sc 12-13-28 PU
1084	850	9-27-06	CaNoR 301	CaNoR 1084	7-18 PK	Sc 5-06-31 PU
1085	851	10-04-06	CaNoR 302	CaNoR 1085	6-14 PK	Sc 10-01-29 PU
1086	852	10-15-06	CaNoR 303	CaNoR 1086	11-15 PK	Sc 8-24-34 PU
1087	853	10-20-06	CaNoR 304	CaNoR 1087	10-17 PK	Sc 6-30-31 PU
1088	854	11-01-06	CaNoR 305	CaNoR 1088	5-14 PK	Sc 8-23-29 PU
1089	855	11-07-06	CaNoR 306	CaNoR 1089	7-15 PK	Sc 11-25-27 PU
1090	856	11-16-06	CaNoR 307	CaNoR 1090	10-14 PK	Sc 4-25-30 PU
1091	857	12-19-06	CaNoR 308	CaNoR 1091	9-14 PK	Sc 8-11-31 PU
1092	858	1-01-07	CaNoR 309*	CaNoR 1092	2-16 PK	Sc 8-22-34 PU
1093	859	1-07-07	CaNoR 310	CaNoR 1093	8-15 PK	Sc 5-13-29 PU
1094	860	1-07-07	CaNoR 311	CaNoR 1094	3-17 PK	Sc 4-25-30 PU
1095	861	1-07-07	CaNoR 312	CaNoR 1095	10-15 PK	Sc 8-23-29 PU
1096	862	1-21-07	CaNoR 313	CaNoR 1096	8-17 PK	Sc 4-25-30 PU
1097	863	2-04-07	CaNoR 314	CaNoR 1097	2-16 PK	Sc 5-13-29 PU
1098	864	2-25-07	CaNoR 315	CaNoR 1098	-nd	Sc 10-09-29 PU
1099	865	5-09-07	CaNoR 316	CaNoR 1099	5-15 PK	Sc 8-22-34 PU
1100	866	3-08-07	CaNoR 317	CaNoR 1100	2-15 PK	Sc 4-25-30 PU
1101	867	3-19-07	CaNoR 318	CaNoR 1101	12-14 PK	Sc 7-16-31 PU
1102	868	4-04-07	CaNoR 319	CaNoR 1102	12-14 PK	Sc 10-04-29 PU

CNR 1083-1102 were ordered by the **Canadian Northern Railway** on November 22nd 1905.

Many locomotives built by Canada Foundry during the first decade of the twentieth century had relatively short lives. The twenty 4-6-0s in the G-11-a class were all scrapped between 1927 and 1934, with an average service life of twenty-four years. This very well may account for the scarcity of photographs of this class. Well worn, although superheated with inside steam pipes to inclined cylinder castings, **1087** was photographed for company records, likely at Transcona in 1931. [*CNR LOCOMOTIVE DATA CARD*]

G-10-b G-11-a

CNR 1103-1104 — 4-6-0 TEN WHEEL TYPE — G-12-a

Specifications							Appliances		Weights	Fuel Capacity		Length	Notes
Cylinder	Gear	Driv.	Pressure	Boiler	T.E.	Haulage	Steam	Stkr.	Drivers/Eng./Total	Water	Coal		
18x26"	S	62"	190#	WT	24000		sat		103/137/235800	5000 gals	10 tons US	60-0'	[orig CRMW]
18x26"	S	57"	190#	WT	23940	23%	sat		104/142/242000	4000 gals	8 tons	60-0'	[CaNoR/CNR]

Schenectady Locomotive Works – ALCO		1904	(S-178)	$15,196		(2) Acquired by CNR 9-01-1919	
	Serial	Shipped	New as	1-1912		Disposition	
			—	G-12-A			
1103	29532	9- -04	H&SW 11	CaNoR 1103		Sc 2- -23 AK	
1104	29533	9- -04	H&SW 12	CaNoR 1104		Sc 5- -25 AK	

CNR 1103 and **1104** were ordered by the **Halifax & South Western Railway** on August 31st 1904. ALCO records show them built in December 1903, thus raising the possibility they were either built for stock or were from a cancelled order. The **Canadian Northern Railway** operated the line for the owners, William Mackenzie, Donald Mann, and R.J. Mackenzie, who had purchased by the line in July 1901. They were renumbered into the **Canadian Northern Railway** system in 1912, and in July 1914 transferred to CaNoR ownership.

CNR 1105-1108 — 4-6-0 TEN WHEEL TYPE — G-13-a

Specifications							Appliances		Weights	Fuel Capacity		Length	Notes
Cylinder	Gear	Driv.	Pressure	Boiler	T.E.	Haulage	Steam	Stkr.	Drivers/Eng./Total	Water	Coal		
20x26"	S	57"	160#	WT	24800	25%	sat		108/143/253000	4500 gals	tons	- '	[CaNoR 1913]
19x26"	S	57"	160#	WT	22400	22%	sat		108/145/258000	5000 gals	10 tons	56-2'	[CNR]

Cooke Locomotive & Machine Company		1901	$15,500			(4) Acquired by CNR 9-01-1919		
	Serial	Shipped	New as	7-1906	1-1912		Disposition	
			—		G-13-A			
1105	2272	4- -01	GNRC 61	CNQ 61	CaNoR 1105		Sc 7-08-25 EH	
1106	2273	4- -01	GNRC 62	CNQ 62	CaNoR 1106		Sc 3-31-25 MV	
1107	2276	5- -01	GNRC 63*	CNQ 63	CaNoR 1107		Sc 8- -25 EH	
1108	2277	5- -01	GNRC 64	CNQ 64	CaNoR 1108		Sc 8- -25 EH	

CNR 1105-1108 were built for the **Great Northern Railway of Canada**. In July 1906, control and operation of the **Mackenzie, Mann & Company Limited** road passed to the **Canadian Northern Quebec Railway**. Renumbered into the **Canadian Northern Railway** system in 1912, they were transferred to CaNoR ownership in July 1914.

GNRC 63 (1107), at Paterson, New Jersey in May 1901, was in a second order for new power for the Quebec road. Unlike the lettering style in the February order from Brooks (see CNR 1343-1346 on page H-25), the road's name was partially written in full in favour of initials on the side of the cab and tender letterboard. There was little danger of confusion with James J. Hill's larger counterpart in the USA, as the two roads were approximately 1260 railway miles apart. [COOKE WORKS PHOTO C-196/ALCO HISTORIC PHOTOS]

CNR 1109   4-6-0 TEN WHEEL TYPE   G-14-a

Cylinder	Gear	Driv.	Pressure	Boiler	T.E.	Haulage	Steam	Stkr.	Drivers/Eng./Total	Water	Coal	Length	Notes
			Specifications				Appliances		Weights	Fuel Capacity		Length	Notes
14&24x26"	S	56"	#		00		sat		/ / 000	gals	tons	- '	[4-6-0C]
19x26"	S	56"	200#	WT	28500	29%	sat		124/151/251000	4000 gals	10 tons	59-7'	[4-6-0]
19x26"	S	57"	200#	WT	27993	28%	H-C		124/151/251000	5000 gals	10 tons	59-7'	[af. 1924]

Baldwin Locomotive Works – Burnham, Williams & Company		1901	$15,500				(1) Acquired by CNR 9-01-1919
	Serial	Shipped	New as	4-6-0	11-1906	1-1912	Superheated	Disposition
			—			G-14-A		
1109	18714	3- -01	QLSJ 18	-nd	MM&Co (CaNoR)–QLSJ 18	CaNoR 1109	7-17 PK	Sc 5-13-33 JD

CNR 1109, built for the **Quebec & Lake St. John Railway**, was acquired with the purchase of the QLSJ by **Mackenzie, Mann & Company Limited**. It was renumbered into the **Canadian Northern Railway** system in 1912 and, in July 1914, transferred to CaNoR ownership. The date of its conversion from a Vauclain compound to simple with inside steam pipes was apparently not recorded.

CNR 1110   4-6-0 TEN WHEEL TYPE   G-15-a

Cylinder	Gear	Driv.	Pressure	Boiler	T.E.	Haulage	Steam	Stkr.	Drivers/Eng./Total	Water	Coal	Length	Notes
			Specifications				Appliances		Weights	Fuel Capacity		Length	Notes
18x24"	S	57"	160#	WE	18600	19%	sat		/120/156750□	4000 gals	tons	- '	[CaNoR 1913]

Canadian Locomotive Company		1902	$15,000			(1) Acquired by CNR 9-01-1919
	Serial	Shipped	New as	4-1909	1-1912	Disposition
			—	—	G-15-A	
1110	577	8-04-02	COR 12*	MM&Co(CaNoR)–COR 12	CaNoR 1110	Sc 9- -25 MV

CNR 1110, ordered in November 1901, was built for the **Central Ontario Railway**. It was acquired in the purchase of the COR by the **Mackenzie, Mann & Company Limited**, renumbered into the **Canadian Northern Railway** system in 1912 as a G-15-a Ten Wheeler. In July 1914, it was transferred to CaNoR ownership.

COR 12 (1110), at Kingston in August 1902, decked out in a GTR style of lettering and numbering, was a single locomotive order for the regional railway.
[CLC PHOTO/QUEEN'S UNIVERSITY ARCHIVES/ DON McQUEEN COLLECTION]

About 1925, COR 12, as **CNR 1110**, was being turned on the "Armstrong" (manual) table at Howland Junction as part of the background for a family "snap" while they waited to change trains. Although not far from COR rails, the location was 35.7 rail miles northeast of Lindsay at a junction of the former Victoria Railway and the Irondale, Bancroft & Ottawa Railway. By that time, CaNoR and CNR alterations included a different pilot, electric headlight, steel-plated cab and rebuilt tender. In order to relocate the bell forward and place it above the check valve, the sand dome was moved further back along the boiler. [AL PATERSON COLLECTION]

CNR 1111-1130					4-6-0 TEN WHEEL TYPE							G-16-a
			Specifications				Appliances		Weights	Fuel Capacity		Length	Notes
Cylinder	Gear	Driv.	Pressure	Boiler	T.E.	Haulage	Steam	Stkr.	Drivers/Eng./Total	Water	Coal		
20x24"	W	57"	170#	EWT	24310	24%	SCH		113/154/277500	5000 gals	10 tons	61-11'	[CaNoR 1913]
20x26"	W	57"	160#	EWT	25785	25%	SCH		113/154/277500	5000 gals	10 tons	61-11'	[CNR 1921]
20x26"	W	57"	160#	EWT	25770	25%	SCH		113/154/277500	5000 gals	10 tons	61-11'	[CNR 1942]
20x26"	W	57"	160#	EWT	25785	25%	SCH		113/154/277500	5000 gals	2700 gals	61-11'	[oil]

Montreal Locomotive Works Ltd. – ALCO   1912   (Q-187) $17,905   (20) Acquired by CNR 9-01-1919

	Serial	Delivered	New as	Stl.	Leased	Mod	To oil	6-1956	8-1957	Disposition	To
			G-16-A	cab	1942=1956			G-16-a	G-16-a		
1111	50809	2- -12	**CaNoR 1111**	12-35 AV		fam	3-52 PK	CNR (1498)/1		Sc 8-07-56 PU	
1112	50810	2- -12	**CaNoR 1112**	2-38 AK		m	1-53 HQ			So 10-24-52 A	**QNSL**
1113	50811	2- -12	**CaNoR 1113**	12-35 AV		tm				Sc 10-05-54 AK	
1114	50812	2- -12	**CaNoR 1114**	7-36 HQ		tm	1-52 PK			Sc 6-04-54 PU	
1115	50813	2- -12	**CaNoR 1115**	11-40 HQ		m				Sc 2-08-54 AK	
1116	50814	2- -12	**CaNoR 1116**						OCS	Sc 12-06-47 PU	
1117	50815	2- -12	**CaNoR 1117**		NAR	f m	4-52 PK	CNR (1499)/1		Sc 6-29-56 PU	
1118	50816	2- -12	**CaNoR 1118**						OCS	Sc 11-29-47 PU	
1119	50817	2- -12	**CaNoR 1119**	4-38 AK		f m		**CNR 1500**/2	CNR 1119	Sc 8-14-61 JB	
1120	50818	2- -12	**CaNoR 1120**	Tender x1160 5-43		m				Sc 9-07-54 AK	
1121	50819	2- -12	**CaNoR 1121**	8-38 AK		m		CNR (1501)/2		Sc 3-04-60 AK	
1122	50820	2- -12	**CaNoR 1122**	12-33 AV		tm				Sc 12-07-54 AK	
1123	50821	3- -12	**CaNoR 1123**	9-34 AK		m	9-53 PK	CNR (1502)/2		Sc 9-13-56 PU	
1124	50822	3- -12	**CaNoR 1124**	10-39 AK		m				Sc 4-18-56 AK	
1125	50823	3- -12	**CaNoR 1125**	10-33 AV		f m	9-53 PK	CNR (1503)/2		Sc 9-27-56 PU	
1126	50824	3- -12	**CaNoR 1126**	10-35 AV		t				Sc 4-18-56 AK	
1127	50825	3- -12	**CaNoR 1127**			t			OCS	Sc 11-29-47 PU	
1128	50826	3- -12	**CaNoR 1128**	11-37 AV		m				Sc 10-05-54 AK	
1129	50827	3- -12	**CaNoR 1129***	12-36 AV		tm		CNR (1504)/2		Sc 3-04-60 AK	
1130	50828	3- -12	**CaNoR 1130**	3-41 AK		m				Sc 8-16-54 AK	

CNR 1111-1160 were built for the **Canadian Northern Railway** in two orders, twenty in Q-187 (CNR 1111-1130) and another thirty in Q-212 (**CNR 1131-1160**). One of the new features noted by the trade press was the outside steam pipes to the cylinders. The *CNR Mechanical Department Locomotive Diagrams* described the main driver of order Q-212 as flanged, ie. not blind – apparently denoting a change from the earlier Q-187. By the late 1930s and early 1940s, steel cabs were being applied to the class. **CaNoR 1116** headed the first train to arrive in Calgary on February 23rd 1914. On May 23rd 1943, both **1144** and **1160** were wrecked in a head-on collision at Mile 18 near Millerville, New Brunswick (between Edmundston & Campbellton). More details can be found in *Canadian Rail*, March 1972. After the accident, the tender of **1160**, identified as G-16 Spare Tender Number 102, went to 1120 in December 1943. **CNR 1120** had been

(text continues on next page)

damaged in a roundhouse fire at Port Tupper, Nova Scotia in 1943. After the retirement of 1120 in 1954, the tender was held for further (unidentified) service until it too, was dismantled at Moncton on August 1st 1958. **CNR 1117** was known to have been leased to the **Northern Alberta Railways** between 1942 and 1956. In 1956, **1119** was re-numbered to clear the block for numbers to be assigned new diesels, but was restored to its original number when the 1100-series was not required. However the cab numerals for the second use of "1119" were painted on the cab, not raised metal as its two previous numbers had been.

In 1952, **1112**, with the steel cab from P-2-a 0-8-0 8010, was sold for $32,000 (after repairs) to Hollinger-Hanna Ltd., agents for the Iron Ore Company of Canada, for operation on the **Quebec North Shore & Labrador Railway** as 1112. In 1962 the QNS&L donated it to the **Canadian Railroad Historical Association** for the Canadi-

Two orders for fifty 4-6-0s from MLW in 1912 and 1913 made the G-16-a class the largest in the group. Just about a decade later, **1156** exhibited the railway's classic style of the 1920s, despite the missing cylinder cover. Unfortunately neither the location in the Maritimes nor the exact date of the photograph is known. At Pictou, Nova Scotia on November 8th 1957, **1138** represented the other end of the time scale. Alterations worth noting, besides the changes in pilot, headlight, bell and whistle location, would be the removal of the casting for the extended piston rod, the addition of the air reservoir mounted on the pilot, single-level running boards with end ladders, and a remodelled cab.
[BOTH: AL PATERSON COLLECTION]

an Railway Museum. In 1993, it was transferred as **CNR 1112** to affiliate **Smiths Falls Railway Museum of Eastern Ontario**. CNR 1158 was sold in 1960 to the **Western Development Museum**, in Saskatoon, Saskatchewan.

(text continues on next page)

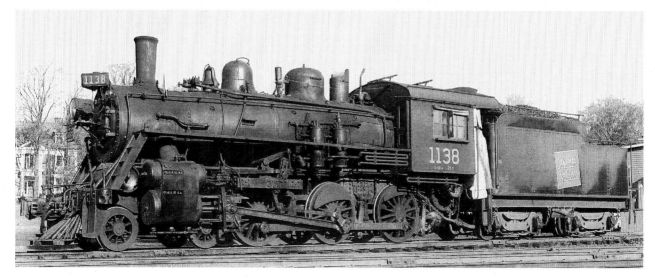

The tenders from **1116**, **1118** and **1127** were set aside as Western Region water transport cars after the engines were retired. Three tenders from the G-16-a class (1131-1160) went into "On Company Service" (OCS) as Atlantic Region auxiliary water tenders after the engines were retired. **CN 102** was used until its retirement in August 1958; **CN 51530** was used with Atlantic Region Building & Bridges 31½-ton crane 50831 [Browning 1941] until after 1962; and **CN 51542** with masonry gang cars until after 1962. The tender of **1136** was listed on December 31st 1955 as being held for possible conversion to an OCS assignment, but no further documentation has been found. The disposition of the tender for **1149**, not scrapped with the engine at Moncton in 1954, is not known.

Two other examples demonstrate how far flung the assignments were for the G-16-a class. White-trimmed and graphited 1111, at Campbellton, New Brunswick on May 28th 1945, even had a cover for the air reservoir located on the pilot.
[AL PATERSON COLLECTION]

One of eight converted to oil, 1158 at Saskatoon on June 2nd 1958, was also amongst many in this class which had cast metal cab numerals applied in the early 1950s. However, the horizontally-mounted tender wafers introduced after 1956 were never used on any G class Ten Wheelers.
[JOHN RIDDELL COLLECTION]

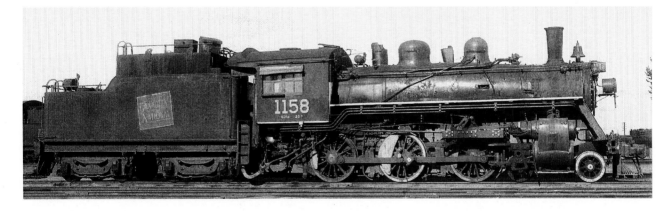

CNR 1131-1160   4-6-0 TEN WHEEL TYPE   G-16-a

Cylinder	Gear	Driv.	Specifications Pressure	Boiler	T.E.	Haulage	Appliances Steam	Stkr.	Weights Drivers/Eng./Total	Fuel Capacity Water	Coal	Length	Notes
20x24"	W	57"	170#	EWT	24310	24%	SCH		113/154/277500	5000 gals	10 tons	61-11'	[CaNoR 1913]
20x26"	W	57"	160#	EWT	25785	25%	SCH		113/154/277500	5000 gals	10 tons	61-11'	[CNR 1921]
20x26"	W	57"	160#	EWT	25770	25%	SCH		113/154/277500	5000 gals	10 tons	61-11'	[CNR 1942]
20x26"	W	57"	160#	EWT	25770	25%	SCH		113/154/277500	5000 gals	2700 gals	61-11'	[oil]

Montreal Locomotive Works Ltd. – ALCO   1913   (Q-212) $19,795   (30) Acquired by CNR 9-01-1919

	Serial	Delivered	New as G-16-A	Stl. cab	Mods	To oil	6-1956 G-16-a	Tender to	Disposition	To
1131	52560	3- -13	CaNoR 1131	4-36 AK	m		CNR (1505)/2		Sc 12-31-56 AK	
1132	52561	3- -13	CaNoR 1132	5-28 HQ	f				Sc 11-22-54 AK	
1133	52562	3- -13	CaNoR 1133	8-28 HQ	m?		CNR (1506)/2		Sc 3-08-57 AK	
1134	52563	3- -13	CaNoR 1134						Sc 10-22-54 AK	
1135	52564	3- -13	CaNoR 1135	1-32 HQ	t		CNR (1507)/2		Sc 5-13-59 MP	
1136	52565	3- -13	CaNoR 1136	2-31 HQ	m			OCS	Sc 7-14-54 C	
1137	52566	3- -13	CaNoR 1137		m?				Sc 2-08-54 AK	
1138	52567	3- -13	CaNoR 1138	-nd	tm		CNR (1508)/2		Sc 8-21-61 JB	
1139	52568	3- -13	CaNoR 1139	6-34 HQ	m		CNR (1509)/2		Sc 3-25-60 AK	
1140	52569	3- -13	CaNoR 1140	11-33 AK	ft	1-52 PK	CNR (1510)/2		Sc 12-31-58 PU	
1141	52570	3- -13	CaNoR 1141	9-34 AV	t				Sc 8-30-55 AK	
1142	52571	3- -13	CaNoR 1142	9-41 MP	tm			CN 51542	Sc 3-02-55 AK	
1143	52572	3- -13	CaNoR 1143		m?				Sc 8-30-55 AK	
1144	52573	3- -13	CaNoR 1144	8-30 HQ	Wr 5-23-43			CN 102	Sc 9-30-43 AK	
1145	52574	3- -13	CaNoR 1145	10-32 HQ	m?		CNR (1511)/2		Sc 9-01-57 AK	
1146	52575	4- -13	CaNoR 1146	7-33 HQ	m?				Sc 11-22-54 AK	
1147	52576	4- -13	CaNoR 1147	11-33 AK	m		CNR (1512)/2		Sc 3-08-57 AK	
1148	52577	4- -13	CaNoR 1148	5-40 AK	m				Sc 3-02-55 AK	
1149	52578	4- -13	CaNoR 1149		tm?				Sc 6-15-54 AK	
1150	52579	4- -13	CaNoR 1150	3-34 HQ	f m	4-52 PK	CNR (1513)/2		Sc 7-27-56 PU	
1151	52580	4- -13	CaNoR 1151	2-39 AK	t				Sc 12-07-54 AK	
1152	52581	4- -13	CaNoR 1152	12-37 AK	tm?		CNR (1514)/2		Sc 2-01-57 AK	
1153	52582	4- -13	CaNoR 1153		m				Sc 11-22-54 AK	
1154	52583	4- -13	CaNoR 1154		m?				Sc 9-07-54 AK	
1155	52584	4- -13	CaNoR 1155		tm?			CN 51530	Sc 7-28-55 AK	
1156	52585	4- -13	CaNoR 1156	10-32 AK	tm?				Sc 12-31-54 AK	
1157	52586	4- -13	CaNoR 1157		m		CNR (1515)/2		Sc 10-04-57 C	
1158	52587	4- -13	CaNoR 1158		ftm	1-52 PK	CNR (1516)/2		Dn 6-03-60 PU	WDM
1159	52588	4- -13	CaNoR 1159	6-40 AK	tm?				Sc 3-09-54 AK	
1160	52589	4- -13	CaNoR 1160	4-33 AV	Wr 5-23-43			CNR 1120	Sc 9-30-43 AK	

CNR 1131-1160: See note under CNR 1111-1130 (page G-18).

CNR 1161-1165 (first)   4-6-0 TEN WHEEL TYPE   G-17-a

Cylinder	Gear	Driv.	Specifications Pressure	Boiler	T.E.	Haulage	Appliances Steam	Stkr.	Weights Drivers/Eng./Total	Fuel Capacity Water	Coal	Length	Notes
22x26"	W	57"	170#	EWT	31800	32%	SCH		133/173/296500	5000 gals	10 tons	63-6½'	[CaNoR 1913]
22x26"	W	58"	180#	EWT	33195	32%	SCH		133/173/297000	5000 gals	10 tons	61-11'	[af. 1924]

Montreal Locomotive Works Ltd. – ALCO   1913   (Q-213) $19,987   (5) Acquired by CNR 9-01-1919

	Serial	Delivered	New as G-17-A	Heavier frame	Stl. cab	Mods	6-1956 G-17-a	Tender to	Disposition
1161	52652	5- -13	CNO–CaNoR 1161			m?		OCS	Sc 3-02-55 AK
1162	52653	5- -13	CNO–CaNoR 1162*		11-26 HQ	tm	CNR (1517)/2		Sc 8-16-57 AK
1163	52654	5- -13	CNO–CaNoR 1163		2-31 HQ	m	CNR (1518)/2		Sc 7-05-57 AK
1164	52655	6- -13	CNO–CaNoR 1164		2-46	t	CNR (1519)/2		Sc 12-27-57 AK
1165/1	52656	6- -13	CNO–CaNoR 1165	3-25 HQ?	6-28 HQ	m			Sc 11-22-54 AK

CNR 1161-1165 were ordered by the **Canadian Northern Railway**, for assignments on the Canadian Northern Ontario Railway, on September 23rd 1912. The remaining twenty-five in Q-213 were assigned **CaNoR** road numbers 1385-1409. By the late 1930s, 1162, 1163 and 1165 had received steel cabs, and by the late 1940s the last, 1161, had been modified. The tender of 1161 was listed on December 31st 1955 as being held for possible conversion to an OCS assignment, but no further information is known.

G-17-a

Photographs from three decades show few structural
changes to the class. **CaNoR 1163**, at an unidentified
location about 1916, still retained its as-built extended
piston rods, cab window arrangement and tender configuration.

CNR 1165, at Quebec City on August 24th 1936,
had been shopped with a boiler tube pilot,
disk pilot wheels, a centred headlight and a steel cab.
[BOTH: AL PATERSON COLLECTION]

At Truro in 1943, **1164** had yet to have its cab replaced, but
had a coal bunker extension and a new location for its bell.

[AL PATERSON COLLECTION]

CNR 1166 (first)   4-6-0 TEN WHEEL TYPE   G-18-a

Cylinder	Gear	Driv.	Specifications Pressure	Boiler	T.E.	Haulage	Appliances Steam	Stkr.	Weights Drivers/Eng./Total	Fuel Capacity Water	Coal	Length	Notes
18x24"	S	57"	170#	WT	20000		sat		80/101/174300	2800 gals	6 tons	53-5'	[orig CGR]
18x24"	S	57"	140#	WT	16200	16%	sat		80/101/187000	3900 gals	7 tons sb?	53-5'	[by 10-1917]

	Serial	Shipped	New as	1-1912	12-1915			Rs	Disposition	To
Canadian Locomotive & Engine Company (Dübs) 1891									(1) Acquired by CNR 9-01-1919	
			—	G6 197%	T1-6 80%					
1166/1	419	10-07-91	**IRC 207**	**IRC 1005**	**CGR 1005**				Sc 2- -23 EH	
—	420	10-13-91	IRC 208	IRC 1006	CGR 1006			sb	Sc 4- -18 AK	ML&P 4
—	421	10-21-91	IRC 209	IRC 1007	CGR 1007			12-18-17	Sc -18 EH	
—	422	10-28-91	IRC 210	IRC 1008	CGR 1008			12-18-17	Sc -18 EH	

CNR 1166 (first) was one of four Ten Wheelers delivered to the **Intercolonial Railway of Canada.** Two were removed from the roster in 1917 and one (CGR 1006) was sold as Maritime Coal, Railway & Power Company, surviving until 1930.

GTW 1167 (first)   4-6-0 TEN WHEEL TYPE   G-19-a

Cylinder	Gear	Driv.	Specifications Pressure	Boiler	T.E.	Haulage	Appliances Steam	Stkr.	Weights Drivers/Eng./Total	Fuel Capacity Water	Coal	Length	Notes
18x24"	S	57"	160#	WT	18553	19%	sat		93/120/190000	gals	tons	55-11'	[CNR]

	Serial	Shipped	New as	1898/1900	11-1904	4-1908	1-1910	Disposition
Baldwin Locomotive Works – Burnham, Parry, Williams & Company 1891 $6502								(1) Acquired by CNR 3-01-1923
					A1	A1	A1	
1167/1	????	-91	**C&GT 152**	**C&GT/GTW 1241**	**GTW 1241**	**GTW 1287**	**GTW 2340**	Sc 11- -23 UB
—	????	-91	C&GT 153	C&GT/GTW 1242	GTW 1242	GTW 1287	GTW 2341	Sc 12- -23 UB

GTW 1167 (first) was one of two 4-6-0s built for the **Chicago & Grand Trunk Railway.** The Baldwin serial numbers are unknown. These were the first 4-6-0s acquired by the GTR. For a summary of the GTR A class, see the CNR I class under 1543-1628 (pages I-12 and I-13). C&GT 153/*GTW* 2341 was scrapped before the road's amalgamation into the CNR.

The oldest pair of GTR 4-6-0s in the CNR G class had been built with extended smokeboxes, evenly-spaced drivers and flared tender letterboards. Short-lived as a GTW Ten Wheeler, **GTR 1241** (1167), at an unidentified location about 1905, had already been shopped with a footboard pilot and tender headlight, suggesting it had been relegated to switching assignments. The design of the tender trucks, common in the 1860s, carried their load on the side bearings rather than on a centre pin. [AL PATERSON COLLECTION]

CNR 1165-1168 (second)   (4) See CNR **1009; 1012; 1017-1018**/2 F-1-b; F-2-a; F-1-c class (pages F-5; F-6; F-8; F-10)

CNR 1168-1177 (first) — 4-6-0 TEN WHEEL TYPE — G-20-a

Cylinder	Gear	Driv.	Pressure	Boiler	T.E.	Haulage	Appliances Steam	Stkr.	Weights Drivers/Eng./Total	Fuel Capacity Water	Coal	Length	Notes
14&24x26"	S	56"	180#		23000		sat		116/145/229000	4000 gals	9 tons	60-0½'	[compound]
19x26"	S	56"	180#	EWT	26000		sat		119/161/298052	5800 gals	9 tons	60-0½'	[GTR A2 simple]
21x26"	S	56"	170#	EWT	29069		SCH		119/161/298052	5800 gals	9 tons	60-0½'	[GTR A7]
21x26"	S	57"	170#	EWT	29069	30%	SCH		119/161/298052	5800 gals	9 tons	60-0½'	[af. 1924]
21x26"	S	57"	170#	EWT	29069	30%	SCH		119/161/298052	5800 gals	9 tons	64-11'	[1174]

Baldwin Locomotive Works – Burnham, Parry, Williams & Company   1898 $17,500   (10) Acquired by CNR 3-01-1923

	Serial	Shipped	New as	6-1899	1-1905	Simple	1-1910	Superheated	Stl cab	Tender to	Disposition	To
			—	H	A2	A2	A2	A7				
1168/1	15912	5- -98	PSCR 629		CAR 629	GTR 1352		GTR 1640	7-15 HQ		Sc 10-01-41 JD	
—	15913	5- -98	OA&PS 630	CAR 630	GTR 1353							**CNR 1177**
1169	15914	5- -98	OA&PS 631*	CAR 631	GTR 1354	by-10	GTR 1641	5-21 HQ	12-15 & nf	Sc 1-13-36 MQ	
1170	15915	5- -98	OA&PS 632	CAR 632	GTR 1355	af-13	GTR 1642	4-22 HQ	OCS	Sc 9-04-41 LM	
1171	15916	5- -98	OA&PS 633	CAR 633	GTR 1356		GTR 1643	4-16 HQ	OCS	Sc 10-01-41 JD	
1172	15917	5- -98	OA&PS 634	CAR 634	GTR 1357	7-14	GTR 1644	4-20 HQ	6-34 HQ	Sc 10-09-35 JD	
1173	15918	5- -98	OA&PS 635	CAR 635	GTR 1358		GTR 1645*	6-15 HQ		Sc 12-28-35 HW	
1174	15919	5- -98	OA&PS 636	CAR 636	GTR 1359*	by-10	GTR 1646	8-19 HQ	CN 51481	Sc 11-18-50 AK	
1175	15920	5- -98	OA&PS 637	CAR 637	GTR 1360	by-10	GTR 1647	1-21 HQ	12-21	Sc 12-30-35 MQ	
1176	15921	5- -98	OA&PS 638	CAR 638	GTR 1361		GTR 1648	12-17 HQ	OCS	Sc 10-01-41 JD	
1177	15913	5- -98	OA&PS 630	CAR 630	GTR 1353	1-10	GTR 1649	1-16 HQ		Sc 2-19-37 MQ	

CNR 1168-1177 were built as Vauclain compounds for the **Canada Atlantic Railway** for use on the **Parry Sound Colonization Railway** or the **Ottawa Arnprior & Parry Sound Railway**. Later, as GTR locomotives (Lot 9), they continued to work in their original territory, even as far east as the CVR at St. Albans, Vermont. **Grand Trunk** rebuilt seven to simple 4-6-0s, and the remaining three were converted to simple engines at the same time as they were superheated. GTR 1353 (1177) was extensively rebuilt in

(text continues on next page)

The only Vauclain compounds in the G class originated with the OA&PS. **OA&PS 631** (1169) on the builder's turntable at Philadelphia in May 1898 appeared much bulkier than it actually was, due in part to the compound cylinders and extended smokebox.
[BLW PHOTO 1054/H.L. BROADBELT/WES DENGATE COLLECTION]

GTR 1640 (1168), with its crew at Swanton Vermont, in September 1914, had yet to be rebuilt simple, but had undergone upgrading modifications to its pilot, headlight, cab window arrangement and coal bunker.
[KARL E. SCHLACHTER PHOTO?/
AL PATERSON & C.A. BUTCHER/WES DENGATE COLLECTIONS]

January 1910 and renumbered at the end of the group as GTR 1649. At the same time, GTR 1642 (**1169**) was superheated, it was fitted with a new design of frame (**nf**). For a summary of the GTR A class, see the CNR I class under 1543-1628 (pages I-12 and I-13).

By the mid 1940s, **1174** had acquired a longer tender (by 4'-11"), and after the engine's retirement, the tender was used as **CN 51481**, with the Charlottetown-assigned 60-ton crane 50142 [Browning 1914], until after 1962. Another four tenders were set aside as Central Region water transport cars after the engines were retired. Although

By the summer of 1944, **1174**, at Belleville, had been altered sufficiently to CNR standards to mask its mechanical origins. The only remaining vestige of its past was (as found on most GTR steamers) maintaining the location of the bell between the sand and steam domes. One of the earliest in the class to have its open wooden cab replaced, **1174** was unique in the class, having acquired a lengthier tender.
[BOTH: JAMES ADAMS PHOTOS/DON McQUEEN COLLECTION]

1170 and **1171** were initially stored at Scarboro pit when they were retired, they were moved to Val Royal in Montreal for dismantling.

CNR 1178 4-6-0 TEN WHEEL TYPE G-21-a

			Specifications				Appliances		Weights		Fuel Capacity		Length	Notes
Cylinder	Gear	Driv.	Pressure	Boiler	T.E.	Haulage	Steam	Stkr.	Drivers/Eng./Total		Water	Coal		
18x26"	S	60"	195#		25000		sat		/ / 000		gals	tons	- '	[orig TH&B]
18x26"	S	57"	195#		24000		sat		/ / 000		gals	tons	- '	[AQ&W]

Schenectady Locomotive Works – ALCO		1907	(S-515)				(1) Acquired by CNR 10-01-1929
	Serial	Shipped	New as	11-1907	10-1925		Disposition
			—	F-2a	—		
1178	44395	11-15-07	**SFRDM 103**	**TH&B 29/2**	**AQ&W 29**		Sc 5-07-37 AK

CNR 1178, ordered originally by the **Santa Fe, Raton & Des Moines Railroad**, was refused and subsequently shipped to the **Toronto Hamilton & Buffalo Railway**. Apparently the TH&B altered the driving wheel diameter from 60 to 57 inches, as later records list them as 57 inches. In 1925, it was sold to the **Atlantic Quebec & Western Railway**. CNR assigned it to trains running between South Devon (Fredericton), McGivney and Newcastle, New Brunswick.

G-21-a

The last G class 4-6-0 to be acquired by CNR had a much travelled career. Shown here at Brantford about 1920 as **TH&B 29**, it remained in Ontario until 1925. It then migrated east, where it was in service for another four years as **AQ&W 29** as seen, likely at New Carlisle, Quebec, about 1926. Seven years later it finished its career as **1178**, and by May 1936, was in the Moncton scrap line.
[ALL: SIRMAN COLLECTION]

1179-1199   **Numbers not used**

CONTENTS

H CLASS: 1200-1454

CNR 1200-1454

H CLASS
4-6-0 TEN WHEEL TYPE

The "H" class was assigned road numbers **1200-1499** for Ten Wheelers with drivers between 58 and 63 inches. The class totalled 259 4-6-0s and came from seven sources: 196 from the **Canadian Northern Railway**; thirty from the **Grand Trunk Pacific Railway**; fourteen from the **Canadian Government Railway** (including three of **Intercolonial Railway of Canada** origin); seven from the **Duluth Winnipeg & Pacific Railway**; six from the **Quebec Montreal & Southern Railway**; four from a CNR "I" class renumbering; and two from the **Quebec Oriental Railway**.

In 1919, the road number allotment assigned **1206-1346**; **1354-1409** to former CaNoR 4-6-0s; **1347-1353** to the ex DW&Ps; and **1410-1423** to those from the CGR. In 1920, numbers **1423-1452** were assigned former GTP Ten Wheelers. In 1929, **1200-1201** were used for a pair of QORs, and **1202-1205**; **1453-1454** were used for those from the QM&S. In 1919, to maintain former CaNoR road number integrity, the numbers **1200-1205** had not been used, and the number **1423** was re-used in 1920 because of an earlier retirement. All dates are mm-dd-yr.

1200-1205 — 4-6-0 TEN WHEEL TYPE

In order to keep the rest of the CaNoR 1200-series numbering and class sequence intact these road numbers were not used in 1919. CaNoR 4-6-0s 1200-1205 are listed below (and note 1206) as a reference.

1200-1201 — 4-6-0 TEN WHEEL TYPE — H-1-A; H-1-B

Cylinder	Gear	Driv.	Pressure	Boiler	T.E.	Haulage	Steam	Stkr.	Drivers/Eng./Total	Water	Coal	Length	Notes
18x24"	S	63"	180#		18884	19%	sat		89/124/196300	5000 gals	tons	- '	[1200 1909]
18x24"	S	63"	180#		16786	17%	sat		89/124/196300	5000 gals	tons	- '	[1201 1909]

Brooks Locomotive Works   1892

	Serial	Shipped	New as	1898	1900	8-01-1902	1-1912	1-1912		Disposition
			—	—	—	—	H-1-A	H-1-B		
	2061	4- -92	ATSF 829	ATSF 593	ATSF 193	CaNoR 58	CaNoR 1200			Sc 8-13-17 PK
	2065	4- -92	ATSF 833	ATSF 507	ATSF 197	CaNoR 59		CaNoR 1201		Sc 6- -17 PK

CaNoR 1200-1201: See 1206 (page H-5).

1202-1206 — 4-6-0 TEN WHEEL TYPE — H-2-A; H-2-B; H-2-C

Cylinder	Gear	Driv.	Pressure	Boiler	T.E.	Haulage	Steam	Stkr.	Drivers/Eng./Total	Water	Coal	Length	Notes
18x24"	S	56"	170#		20056		sat		95/127/212000	4100 gals	tons	- '	[1202 & 1205 1909]
18x24"	S	62"	170#		18123		sat		95/127/212000	4100 gals	tons	- '	[1203 & 1204 1909]
18x24"	S	62"	180#		17000	17%	sat		95/127/212000	4100 gals	tons	- '	[H-2-A]
18x24"	S	62"	180#		20056	20%	sat		95/127/212000	4100 gals	tons	- '	[H-2-B]

Baldwin Locomotive Works – Burnham, Williams & Company   1894

	Serial	Shipped	New as	1898	1900	Sold	to	H-2-A	H-2-B	H-2-C	Disposition
			—	—	—	1902			1-1912		
	13957	3- -94	ATSF 856	ATSF 641	ATSF 260	8-01-02	CaNoR 62			CaNoR 1206	to **CNR 1206**
	13958	3- -94	ATSF 857	ATSF 642	ATSF 256	8-01-02	CaNoR 60	CaNoR 1203			Sc 8-18-17 PK
	13959	3- -94	ATSF 858	ATSF 638	ATSF 257	8-01-02	CaNoR 61	CaNoR 1204			Sc 5-10-17 PK
	13960	3- -94	ATSF 859	ATSF 639	ATSF 258	6-30-02	CaNoR 56	CaNoR 1202			Sc 7-07-17 PK
	13961	3- -94	ATSF 860	ATSF 640	ATSF 259	6-30-02	CaNoR 57		CaNoR 1205		Sc 8-18-17 EM

CaNoR 1202-1206: See 1206 (page H-5).

CNR 1200 — 4-6-0 TEN WHEEL TYPE — H-1-a

Cylinder	Gear	Driv.	Pressure	Boiler	T.E.	Haulage	Steam	Stkr.	Drivers/Eng./Total	Water	Coal	Length	Notes
18x24"	S	62"	165#		00	19%	sat		95/119/154255	3000 gals	10 tons	57-5'	[CPR SR, D3ʙ 1907]
18x24"	S	63"	165#		00	17%	sat		/ / 000	gals	tons	- '	[CPR D3ʙ 1913]

Canadian Pacific Railway – New Shops (DeLorimier Ave.), Montreal   1892   (1) Acquired by CNR 10-01-1929

	Serial	Shipped	New as	4-1907	3-1913	9-1917	nd	Disposition
			SR	D3ʙ	D3ʙ	—	—	
1200	1188	12- -92	**CPR 619**	**CPR 420**	**CPR 7320**	QOR 7	QOR 16	Sc -29 C

CNR **1200** and **1201** were from a group of thirty-six saturated 4-6-0s built between 1892 and 1894 by the **Canadian Pacific Railway** as I: 585=630 under boiler numbers #1157=1199. In 1907, the group was renumbered to II: 395=433, and again in 1912-13 to III: 7295=7329. For specific renumbering see Lavallée: *Canadian Pacific Steam Locomotives*. In 1917, the two were sold by the CPR to the **Quebec Oriental Railway**. Although QOR 16 was assigned a CNR road number and class, it was not "taken into records".

There were two locomotives from predecessor companies which could have possibly become CNR class H-1-a 1200. In both cases the number and class were never applied. The first was Canadian Northern H-1-A 1200, scrapped just before the creation of the CNR. This number (and eight others in the series) were left open to avoid renumbering the rest of the 1200, 1300 and 1400s from Canadian Northern, the owner of the majority of the H class Ten Wheelers. **QOR 16**, at an unidentified location sometime in the 1920s, was the second of the two which might have become CNR H-1-a 1200, had it not been sent to scrap unrenumbered and unclassed the same year it was acquired. [JOSEPH E. PLATT PHOTO/AL PATERSON COLLECTION]

CNR 1201					4-6-0 TEN WHEEL TYPE							H-1-b
			Specifications			Appliances		Weights	Fuel Capacity		Length	Notes
Cylinder	Gear	Driv.	Pressure	Boiler	T.E.	Haulage	Steam	Stkr.	Drivers/Eng./Total	Water	Coal	
18x24"	S	62"	165#		00	19%	sat		95/119/154255	3000 gals	10 tons	57-5'	[CPR SR, D3B 1907]
18x24"	S	63"	165#		00	17%	sat		/ / 000	gals	tons	- '	[CPR D3B 1913]

Canadian Pacific Railway – New Shops (DeLorimier Ave.), Montreal			1892					(1) Acquired by CNR 10-01-1929
	Serial	Shipped	New as	4-1907	12-1912	11-1917	nd	Disposition
			SR	D	D3B	—	—	
1201	1182	10- -92	**CPR 613**	**CPR 415**	**CPR 7315**	**QOR 6**	**QOR 17**	Sc 6-30-30 AK

CNR 1201: see note under CNR 1200 (above).

QOR 17, assigned but never renumbered or classed to CNR H-1-b 1201, was also at an unidentified location sometime in the 1920s. Its CPR lineage, as found with QOR 16 (above), was most noticeable in the style of wooden cab, and by the shape of steam and sand domes. Slight physical differences likely accounted for the CNR Mechanical Department assigning the two 4-6-0s to separate sub-classes, despite both QOR 16 and 17 being built in the same builder's lot. Type of headlight casings, dome, bell, check valve, tool box placement, (QOR 16's bell was immediately in front of the cab) and smokebox steps were some of the more subtle differences. [JOSEPH E. PLATT PHOTO/AL PATERSON COLLECTION]

CNR 1202-1205 (first, second) 4-6-0 TEN WHEEL TYPE H-2-a

Specifications						Appliances		Weights	Fuel Capacity		Length	Notes
Cylinder	Gear	Driv.	Pressure	Boiler	T.E.	Haulage Steam	Stkr.	Drivers/Eng./Total	Water	Coal		
21x26"	S	63"	200#	Wootton	28000	sat		/132/175000	gals	tons	- '	[orig]
22x26"	S	72"	225#	Wootton	34000	SCH		/157/204800	gals	tons	- '	[1203]

Locomotive & Machine Company of Montreal	1907	(Q-74)					(4) Acquired by CNR 7-16-1929
	Serial	Shipped	New as	1910	1919	Superheated	Disposition
			D-3	D-3	D-3		
1202/1	43702	8- -07	**(D&H)–QM&S 200**				Sc 12-31-30 HQ
—	43703	8- -07	(D&H) – QM&S 201	NJR 201	D&H 560		
—	43704	8- -07	(D&H) – QM&S 202	NJR 202	D&H 535		
1203/2	43705	8- -07	**(D&H)–QM&S 203**			9-25 MLW	Sc 6-30-31 HQ
1204/2	43706	8- -07	**(D&H)–QM&S 204**				Sc 4-29-31 JD
1205/2	43707	8- -07	**(D&H)–QM&S 205**				Sc 4-29-31 HQ
—	43708	8- -07	(D&H) – QM&S 206	NJR 206	D&H 561		
—	43709	8- -07	(D&H) – QM&S 207	NJR 207			

The proposed first use of road number 1202 and the second use of 1203 to 1205 was never applied. Both **QM&S 204** and **QM&S 205** (assigned CNR 2nd 1204 and 1205 respectively), at Sorel in 1929, were dismantled as QM&S Ten Wheelers. Although relatively unaltered by the end of their service careers, QM&S 205 carried the last lettering variation stencilled on its cab and tender. The earlier initials similar to those of QM&S 204 were visible under the fading coat of paint on the 205's tender letterboard.
[BOTH: *CNR LOCOMOTIVE DATA CARD*]

CNR 1202-1205 were from eight 4-6-0s ordered under Q-74 for the **Quebec Montreal & Southern Railway** by the **Delaware & Hudson Corporation.** The remaining four were transferred to the D&H-owned Napierville Junction Railway in 1910. In 1919, all but NJR 207 were transferred to parent D&H and were not scrapped until either the early 1940s or the early 1950s. If the four had seen active CNR service, they would have been the only steamers with a Wootton firebox on the system.

CNR 1203-1206 (first, second) 4-6-0 TEN WHEEL TYPE second H-2-c

Renumbered and reclassed from CNR I-6-a (first) 1543-1546 in 1920.

(4) See CNR **1543-1546** I-6-a class (page I-11).

CNR 1206 (first) 4-6-0 TEN WHEEL TYPE first H-2-c

Cylinder	Gear	Driv.	Specifications Pressure	Boiler	T.E.	Haulage	Appliances Steam	Stkr.	Weights Drivers/Eng./Total	Fuel Capacity Water	Coal	Length	Notes
18x24"	S	62"	150#		15991	15%	sat		95/127/212000	4100 gals	tons	- '	[CaNoR 1909]
18x24"	S	56"	180#		21245	19%	sat		95/127/212000	4100 gals	tons	- '	[H-2-C]

Baldwin Locomotive Works – Burnham, Williams & Company			1894					(1) Acquired by CNR 9-01-1919	
	Serial	Shipped	New as	1898	1900	8-01-1902	1-1912		Disposition
			—	—	—	—	H-2-C		
1206/1	13957	3- -94	**ATSF 856**	**ATSF 641**	**ATSF 260**	**CaNoR 62**	**CaNoR 1206**		Sc 10-03-19 ET

CNR 1206 was the only one of seven 4-6-0s to become CNR power which were purchased second-hand in May 1902 by the **Canadian Northern Railway** from the Atchison Topeka & Santa Fe Railway through dealer **James T. Gardiner**. Five of these, including **1206**, had been part of an order from Baldwin built as **ATSF 856-860**. The backgrounds of the five not taken into CNR stock can be found in the roster under **(CaNoR) 1202-1205** on page H-2.

The pair of Brooks-built Ten Wheelers included in this May 1902 sale through **James T. Gardiner** eventually became **CaNoR 1200-1201**, and were from twenty originally built as ATSF 814-833 (page H-2).

CNR 1207-1208 4-6-0 TEN WHEEL TYPE H-3-a

Cylinder	Gear	Driv.	Specifications Pressure	Boiler	T.E.	Haulage	Appliances Steam	Stkr.	Weights Drivers/Eng./Total	Fuel Capacity Water	Coal	Length	Notes
18x24"	S	63"	200#	EWT	20983	21%	sat		104/138/236300	4500 gals	9 tons	64-2'	[CaNoR 1909]
18x24"	S	63"	200#	EWT	20983	21%	H-C		104/138/236300	4500 gals	9 tons	64-2'	
18x24"	S	63"	200#	EWT	20983	21%	H-C		104/138/240600	4500 gals	8½ tons	60-0'	[1207 G-10-b tender]

Brooks Locomotive Works – ALCO		1903	(B-895)	$12,850					(2) Acquired by CNR 9-01-1919	
	Serial	Shipped	New as	1-1912	Superheated	Stl cab	Wooden Hopper	Mods	Tender to	Disposition
			—	H-3-A						
1207	27214	6- -03	**CaNoR 67**	**CaNoR 1207**	11-14 PK	11-35 MP	11-43	m		Sc 12-02-55 LM
1208	27215	6- -03	**CaNoR 68***	**CaNoR 1208**	10-17 PK	11-40 MP	3-47	m	OCS	Sc 4-28-54 LM

CNR 1207 and **1208** were ordered by the Imperial Rolling Stock Company for the **Canadian Northern Railway** on July 9th 1902, and put into service during December 1903. At the same time another twenty-seven Ten Wheelers (1058-1082 and 1209-1210) were ordered. Both **1207** and **1208** were equipped with steel cabs after 1935, and **1207** acquired a tender from a retired G-10-b class 4-6-0. The tender of **1208** was listed on December 31st 1955

Canadian Northern placed two small orders for nearly identical 63-inch driver 4-6-0s from two ALCO affiliates. **CaNoR 68** (1208), at Dunkirk, New York in June 1903, was one of the imports which was one ton heavier and four feet longer than its Canadian counterparts.
[BROOKS WORKS PHOTO B-40/GEORGE CARPENTER COLLECTION]

as held for possible conversion to an OCS assignment, but no further information has been found.

CaNoR 68, almost forty years later as **CNR 1208**, was at Palmerston in August 1941. Although CNR had modified its front end, added running board ladders and a steel cab, it retained its inside steam pipes, inclined cylinders, tender and Brooks builder's plate.
[JAMES ADAMS PHOTO/AL PATERSON COLLECTION]

CNR 1207, at Kincardine a little later in the same year on October 4th, had also undergone similar modifications and retentions, except for the disappearance of its builder's plate and tender being replaced by one from a G-10-b Ten Wheeler. [GEORGE HARRIS PHOTO/ LAWRENCE A. STUCKEY/WES DENGATE COLLECTION]

CNR 1209-1210 — 4-6-0 TEN WHEEL TYPE — H-3-a

Specifications							Appliances		Weights	Fuel Capacity		Length	Notes
Cylinder	Gear	Driv.	Pressure	Boiler	T.E.	Haulage	Steam	Stkr.	Drivers/Eng./Total	Water	Coal		
18x24"	S	63"	200#	EWT	21321	21%	sat		102/136/236300	4500 gals	tons	- '	[CaNoR 1903]
19x24"	S	63"	200#	EWT	23379	23>21%	H-C		102/136/234300	4500 gals	9 tons	60-1½'	

Locomotive & Machine Company of Montreal		1904	(Q-12)	$14,108		(2) Acquired by CNR 9-01-1919	
	Serial	Shipped	New as	1-1912	Superheated	Disposition	
			—	H-3-A			
1209	30143	10-03-04	**CaNoR 69***	**CaNoR 1209**	4-16 PK	Sc 1-31-36 MQ	
1210	30144	10-03-04	**CaNoR 70**	**CaNoR 1210**	2-17 PK	Sc 12-23-35 LM	

CNR 1209 and **1210** were ordered on May 20th 1904 by Imperial Rolling Stock Company for the **Canadian Northern Railway**. At the same time, another twenty-seven Ten Wheelers (1058-1082, 1207 and 1208) were ordered. The change in cylinder bore was made when the engines were superheated. The haulage rating was altered to 21% by 1930 to give a uniform rating for the H-3 class.

Despite some subtle differences (see CaNoR 68 page H-5), **CaNoR 69** (1209) at Montreal in October 1904 varied only physically in boiler jacket and builder's plate style.

Despite these basic similarities, the MLW pair were removed from the roster twenty years before the duo from Dunkirk.
[MLW PHOTO Q-12/WES DENGATE COLLECTION]

CNR 1211-1220 — 4-6-0 TEN WHEEL TYPE — H-3-b

Cylinder	Gear	Driv.	Pressure	Boiler	T.E.	Haulage	Steam	Stkr.	Drivers/Eng./Total	Water	Coal	Length	Notes
			Specifications				Appliances		Weights			Length	Notes
18x24"	S	62"	200#	EWT	21321	21%	sat pv		116/142/262000	5000 gals	10 tons	60-7'	[orig]
18x24"	S	63"	200#	EWT	20983	21%	H-C		116/142/262000	5000 gals	10 tons	60-7'	[1914]
18x24"	S	63"	200#	EWT	20985	21%	H-C		116/142/262000	5000 gals	10 tons	60-7'	[1950]

Locomotive & Machine Company of Montreal   1906   (Q-36) $16,862   (9) Acquired by CNR 9-01-1919

	Serial	Delivered	New as	Orig.	1-1912	Superheated	Stl cab	Tender	Disposition	To
			—	assigned	H-3-B			to		
1211	39782	5-15-06	MMCo/**CaNoR 71**		**CaNoR 1211**	9-14 PK			Sc 2-13-37 MQ	
1212	39783	5-15-06	MMCo/**CaNoR 72**		**CaNoR 1212**	12-14 PK	5-41 MP	OCS	Sc 12-19-50 LM	
1213	39786	5-23-06	MMCo/**CaNoR 73**		**CaNoR 1213**	5-15 PK			Sc 12-03-35 LM	
1214	39787	5-23-06	MMCo/**CaNoR 74**		**CaNoR 1214**	2-14 PK			Sc 8-01-41 MQ	
1215	39790	5-23-06	MMCo/**CaNoR 75**	H&SW	**CaNoR 1215**	10-14 PK		OCS	Sc 12-16-35 AK	
1216	39791	5-23-06	MMCo/**CaNoR 76***	H&SW	**CaNoR 1216**	7-13 PK	7-34 AK		Sc 9-29-39 AK	
1217	39784	5-17-06	MMCo/**CaNoR 77**	H&SW	**CaNoR 1217**	1-21 AK	5-34 AK	CN 1217	Sc 9-09-50 AK	
1218	39785	5-17-06	MMCo/**CaNoR 78**	H&SW	**CaNoR 1218**	5-20 AK	11-36 AK	OCS	Sc 11-21-47 AK	
—	39788	5-29-06	MMCo/CaNoR 79		CaNoR 1219				So -17 A	IR&C 1
1220	39789	6-01-06	MMCo/**CaNoR 80**		**CaNoR 1220**	2-24 AK			Sc 5-03-39 AK	

Superheated **CaNoR 78** (1218), with inclined cylinders and inside steam pipes, was on the H&SW at Bridgewater, Nova Scotia in 1912.
[DETAIL FROM CNR PHOTO 148-5/ WES DENGATE COLLECTION]

CNR 1211-1220 were ordered by the **MacKenzie, Mann & Company** but lettered and numbered into the **Canadian Northern Railway** system. Four were intially assigned to the **Halifax & Southwestern Railway**.

In 1917 **CaNoR 1219** was sold to the **Inverness Railway & Coal Company** as IR&C second 1. The Ten Wheeler had an interesting service career. As **CaNoR 79**, although originally assigned to the Halifax & South Western Railway, it was leased to IR&C in 1908 for use on its mixed train. It was extensively damaged in a derailment at Glendyre (13.5 miles from Inverness, Nova Scotia) on July 11th 1912. Repaired, and given the CaNoR 1912 series number of **1219**, it remained lettered "Canadian Northern". In 1917, it again sustained heavy damage in a wreck at Craigmore, Nova Scotia. When the IR&C assets were sold by Mackenzie & Mann in 1917 to divorce the Cape Breton Railway from the CaNoR system, **1219** was shown in the CaNoR record as "wrecked beyond repair" – hence the justification to sell it to the railway rather than return it to CaNoR stock. IR&C subsequently repaired the locomotive as IR&C second 1 and it was still on the roster on August 7th 1929 when IR&C was acquired by the CNR.

CaNoR 78, thirty-two years later, as **CNR 1218**, at Truro in February 1944, it had been rebuilt with a steel cab but had yet to acquire a centred headlight and running board ladders.
[AL PATERSON COLLECTION]

CNR scrapped it in 1929 without taking it into stock. For more detail of the Glendyre wreck, see McBean: "Derail on the Glendyre Grade", (illus.) in the *UCRS Newsletter* 9-1970 p. 13.

CNR 1214 was stored in the Scarboro pit between 1937 and 1940. The tender of **1217** was used as an OCS auxiliary tender for the Atlantic Region Rail Grinding Gang until sometime after 1962. The disposition of the tenders from **1212**, **1215** and **1218**, not scrapped with the engine at Moncton in 1935, is not known.

CNR 1212, at Palmerston in 1947, had undergone similar alterations to those made on 1218 (above). However, additional changes included spoked pilot wheels, a different bell location and a cut-down tender.
[AL PATERSON COLLECTION]

CNR 1221-1230   4-6-0 TEN WHEEL TYPE   H-4-a

Cylinder	Gear	Driv.	Specifications Pressure	Boiler	T.E.	Haulage	Appliances Steam	Stkr.	Weights Drivers/Eng./Total	Fuel Capacity Water	Coal	Length	Notes
18x24"	S	63"	200#	WT	20983		sat pv		107/135/257000	5000 gals	tons	- '	[CaNoR 1909]
18x24"	S	63"	200#	WT	20980	21%	H-C		109/137/252820	5000 gals	10 tons	59-2'	
19x24"	S	63"	200#	WT	23793	23>21%	H-C		109/137/252820	5000 gals	10 tons	59-2'	

	Serial	Shipped	New as	1-1912	Superheated	19x24"	Stl cab	Mods	8-02-56	Mods	Disposition	To
Canadian Locomotive Company		1906	$16,227								(10) Acquired by CNR 9-01-1919	
				H-4-A								
1221	736	8-16-06	CaNoR 81	CaNoR 1221	1-18 PK						Sc 1-15-36 HW	
1222	737	8-22-06	CaNoR 82	CaNoR 1222	2-17 PK	5-28 MP					Sc 6-26-36 MQ	
1223	738	8-27-06	CaNoR 83	CaNoR 1223	12-15 PK	6-26 HQ	6-34 HQ	m	CNR 1520/2	m	So 9-15-60 C	CRHA
1224	739	9-03-06	CaNoR 84	CaNoR 1224	10-19 PK		10-34 HQ	m?			Sc 3-01-54 LM	
1225	740	9-05-06	CaNoR 85	CaNoR 1225	1-17 PK	1-17 PK					Sc 11-15-35 PU	
1226	741	9-12-06	CaNoR 86	CaNoR 1226	10-17 PK		12-34 HQ?				Sc 12-09-35 HW	
1227	742	9-15-06	CaNoR 87*	CaNoR 1227	7-20 PK	7-20 PK	9-39 MP				Sc 11-28-51 LM	
1228	743	9-20-06	CaNoR 88	CaNoR 1228	3-18 PK						Sc 11-23-35 HW	
1229	744	9-26-06	CaNoR 89	CaNoR 1229	11-17 PK						Sc 11-28-41 LM	
1230	745	10-04-06	CaNoR 90	CaNoR 1230	3-17 PK						Sc 11-15-35 PU	

CNR 1221-1230 were ordered on December 7th 1905 by the **Canadian Northern Railway**. Except for 1222, the change in cylinder bore took place when the engines were superheated. After 1930, those with a 23% haulage rating were changed to 21% to make them uniform with the rest of the H-4 class.

    CNR 1223 was renumbered to clear the 1200-series for the renumbered GR-12e SW1200RS diesel switchers.

(text continues on next page)

CaNoR 87 (1227), new at Kingston in September 1904, was built with piston valves, inclined cylinders and inside steam pipes. The headlights of the class would not be relocated to the centre of the smokebox until the 1920s, and the step on the steam chest casing would be replaced by two-step running board ladders during the superheating cycle… [CLC, MILN-BINGHAM LITHOGRAPH FROM A HENDERSON PHOTOGRAPH/ DON McQUEEN COLLECTION] but by October 16th 1941, **1224**, at Palmerston, superheated with outside steam pipes, had acquired the standard CNR boiler tube pilot, centred headlight and full running board ladders. At least four others in the class had the bell mounted ahead of the stack. [EARL A. ELLIOTT/DON McQUEEN COLLECTION]

In 1960 it was sold for $3140 to the **Canadian Railroad Historical Association** for the **Canadian Railway Museum** at St. Constant, Quebec. In 1991, it was transferred to the **Central British Columbia Railway & Forest Industry Museum** at Prince George, British Columbia.

CNR 1231-1245 — 4-6-0 TEN WHEEL TYPE — H-4-b

Cylinder	Gear	Driv.	Pressure	Boiler	T.E.	Haulage	Steam	Stkr.	Drivers/Eng./Total	Water	Coal	Length	Notes
18x24"	S	63"	200#	WT	20983		sat pv		107/135/257000	5000 gals	tons	- '	[CaNoR 1913]
18x24"	S	63"	200#	WT	20980	21%	H-C		109/137/252820	5000 gals	10 tons	50-9'	
19x24"	S	63"	200#	WT	23793	23>21%	H-C		109/137/252820	5000 gals	10 tons	50-9'	

Canadian Locomotive Company 1907 $17,380 (15) Acquired by CNR 9-01-1919

	Serial	Shipped	New as			1-1912	Superheated	19x24"	Stl cab	Disposition
				—	—	—	H-4-B			
1231	751	4-26-07	**CNO 171**			CaNoR	**1231**	12-16 PK			Sc 4-25-30 PU
1232	752	5-01-07	**CNO 172**			CaNoR	**1232**	1-17 PK			Sc 12-23-35 LM
1233	753	5-08-07	**CNO 173**			CaNoR	**1233**	5-17 PK	5-17 PK		Sc 5-25-37 PU
1234	754	5-14-07	**CNO 174**			CaNoR	**1234**	2-17 PK	2-17 PK		Sc 5-25-37 PU
1235	755	5-18-07	**CNO 175**			CaNoR	**1235**	8-19 PK			Sc 2-28-36 LM
1236	756	5-23-07	**CNO 176**			CaNoR	**1236**	5-21 MV		5-40 MP	Sc 3-01-54 LM
1237	757	5-29-07	**CNO 177**			CaNoR	**1237**	10-17 PK	10-17 PK		Sc 10-07-35 PU
1238	758	6-04-07	**CNO 178**			CaNoR	**1238**	2-17 GV		6-36 MP	Sc 4-28-54 LM
1239	759	6-09-07	**CNO 179**			CaNoR	**1239**	8-19 PK			Sc 11-22-35 PU
1240	760	6-13-07	**CNO 180**			CaNoR	**1240**	10-18 PK	10-18 PK		Sc 6-30-36 PU
1241	761	6-20-07		**HSW 181**		(CaNoR) HSW	**1241**	11-21 AK			Sc 12-31-41 AK
1242	762	6-25-07		**HSW 182**		(CaNoR) HSW	**1242**	6-23 AK			Sc 8-31-39 AK
1243	763	6-29-07			**CNQ 183**	CaNoR	**1243**	5-17 PK			Sc 5-25-37 PU
1244	764	7-08-07			**CNQ 184**	CaNoR	**1244**	9-17 PK	9-17 PK	1-34 MP	Sc 11-28-51 LM
1245	765	7-13-07	**CNO 185**			CaNoR	**1245**	1-17 PK			Sc 6-30-36 PU

CNR 1231-1245 were ordered on June 29th 1906 by parent **Canadian Northern Railway** for three of its subsidiary lines. **Canadian Northern Ontario Railway** were assigned eleven, the **Halifax & South Western Railway** two, and two went to the **Canadian Northern Quebec Railway**. The change in cylinder bore took place when the engines were superheated. After 1930, those with a 23% haulage rating were changed to 21% to make them uniform with the rest of the H-4 class.

Although lettered for three Canadian Northern subsidiaries, all were built with piston valves, inclined cylinders and inside steam pipes. Lettered Canadian Northern Ontario and thought to be on the South Parry turntable between 1910 and 1912, **171** (1231) had already been fitted with a turbo-generator and electric headlight installed inside the older casing.
[H.L. GOLDSMITH/GEORGE CARPENTER COLLECTION]

At Bridgewater sometime between 1907 and 1911, 182 (1242) had yet to lose its kerosene headlight or tender's flared letterboard and was still lettered Halifax & South Western.
[NMS&T PHOTO 4433/ DON McQUEEN COLLECTION]
Despite the slight variance of class light position, both Ten Wheelers (171 and 182) had steel sheathed cabs, vertical stave pilots, bolted disk pilot wheels and upper quadrant smokebox handrails.

Under CNR ownership, changes in appearances continued. **CNR 1244**, at Palmerston in 1939, had the usual boiler tube pilot and steel cab alterations, as well as being modified with outside steam pipes, full running board ladders and spoked pilot wheels.

As with most of the sub-class, 1244 retained its headlight in front of the stack and the bell behind it.
[LAWRENCE A. STUCKEY/WES DENGATE COLLECTION]

CNR 1238, at Truro on July 8th 1945, had acquired a steel cab, but kept the as-built hanging running board steps and inside steam pipes. By the late 1930s, however, the headlight had been centred and the bell moved to the headlight's as-delivered position.
[ERNIE PLANT PHOTO/ WES DENGATE COLLECTION]

H-4-b

CNR 1246-1260   4-6-0 TEN WHEEL TYPE   H-5-a

Cylinder	Gear	Driv.	Specifications Pressure	Boiler	T.E.	Haulage	Appliances Steam	Stkr.	Weights Drivers/Eng./Total	Fuel Capacity Water	Coal	Length	Notes
19x26"	S	63"	200#	WT	25326	26%	sat pv		131/164/295250	6000 gals	10 tons	- '	[1909-1913]
19x26"	S	63"	200#	WT	25400	24%	H-C		116/164/278100	5000 gals	10 tons	64-3½'	

Canada Foundry Company Ltd.   1907 $18,830   (15) Acquired by CNR 9-01-1919

	Serial	Shipped	New as	1-1912	Superheated	Disposition
			—	H-5-A		
1246	869	8-17-07	CaNoR 321	CaNoR 1246	1-16 PK	Sc 5-31-37 PU
1247	870	8-17-07	CaNoR 322	CaNoR 1247	7-16 PK	Sc 9-30-39 PU
1248	871	8-23-07	CaNoR 323	CaNoR 1248	11-15 PK	Sc 5-25-37 PU
1249	872	8-30-07	CaNoR 324	CaNoR 1249	4-16 PK	Sc 7-09-31 PU
1250	873	8-30-07	CaNoR 325	CaNoR 1250	4-17 PK	Sc 5-25-37 PU
1251	874	9-04-07	CaNoR 326	CaNoR 1251	5-15 PK	Sc 10-14-29 PU
1252	875	9-09-07	CaNoR 327	CaNoR 1252	6-17 PK	Sc 9-30-39 PU
1253	876	9-14-07	CaNoR 328	CaNoR 1253	1-17 PK	Sc 9-13-35 PU
1254	877	9-20-07	CaNoR 329	CaNoR 1254	11-15 PK	Sc 12-19-35 PU
1255	878	9-22-07	CaNoR 330	CaNoR 1255	7-15 PK	Sc 12-19-35 PU
1256	879	10-10-07	CaNoR 331	CaNoR 1256	5-15 PK	Sc 11-22-35 PU
1257	880	10-10-07	CaNoR 332	CaNoR 1257	8-15 PK	Sc 6-23-31 PU
1258	881	10-12-07	CaNoR 333	CaNoR 1258	7-15 PK	Sc 9-18-28 PU
1259	882	10-16-07	CaNoR 334	CaNoR 1259	10-15 PK	Sc 10-16-29 PU
1260	883	10-19-07	CaNoR 335	CaNoR 1260	7-16 PK	Sc 8-28-44 PU

Many main line steamers built by Canada Foundry had relatively short service lives. Most of the H-5-a class were removed from the roster in the mid- to late-1930s. **CNR 1251**, believed to be retired, and at Transcona in 1929, still had a horizontal slatted pilot and tender lettering, but its electric headlight had been removed for reinstallation on another steamer.
[*CNR LOCOMOTIVE DATA CARD*]

CNR 1246-1260 were ordered on December 15th 1906 by the **Canadian Northern Railway**. CaNoR 1260 was assigned CaNoR "East Lines" until April 24th 1923. These lines, including the Quebec & Lake St. John Railway and Halifax & South Western Railway, were all east of Montreal. Although **CaNoR 335** (1260) was wrecked at St. Prosper, Quebec on October 22nd 1908, it was repaired and returned to service on November 29th 1909.

CNR 1257, also believed retired and destined for the scrap line at Transcona in 1931, now with a slatted pilot and tender lettering, its electric headlight in the former kerosene casing. The hanging running board ladders and inclined cylinders with inside pipes were as-built

configurations. In both cases, the rectangular builder's plate with notched corners mounted on the cylinder casing had survived a quarter century of shopping and exposure to the elements. [*CNR LOCOMOTIVE DATA CARD*]

CNR 1261-1262 — 4-6-0 TEN WHEEL TYPE — H-6-a

Cylinder	Gear	Driv.	Pressure	Boiler	T.E.	Haulage	Steam	Stkr.	Drivers/Eng./Total	Water	Coal	Length	Notes
20x26"	S	63"	180#	BEL	28060	28%	sat pv		126/157/277000	5000 gals	10 tons	- '	[CaNoR 1909]
20x26"	S	63"	200#	BEL	28060	28%	H-C/SCH		126/157/274500	5000 gals	10 tons	61-1'	

Brooks Locomotive Works – ALCO   1903   (B-892) $13,300   (2) Acquired by CNR 9-01-1919

	Serial	Shipped	Ordered as	New as	7-1906	1-1912	—— Superheated ——	Disposition
			—	—	—	H-6-A		
—	26865	10- -02	(GNRC 69)?					To PM 198
—	26866	10- -02	GNRC 70*					To PM 199
1261	26880	6- -03	(GNRC 71)?	GNRC 200	CaNoR–CNQ 200	CaNoR 1261	9-18 PK	Sc 3-17-45 PU
1262	26881	6- -03	(GNRC 72)?	GNRC 201	CaNoR–CNQ 201	CaNoR 1262	6-18 PK 2-30 PK	Sc 3-17-45 PU

CNR 1261 and 1262 were in an order placed by the Canadian Northern Railway for the Great Northern Railway of Canada on May 16th 1902. However a mystery surrounds the quantity built and the road numbers used by the GNRC. Brooks records ("Dunkirk" on *CNR Mechanical Department Locomotive Diagrams*) show the company built two orders of 4-6-0s with Belpaire fireboxes in late 1902. The first, B-891, using serials #26865-26866, was shipped in October 1902 as Pere Marquette Railroad 198 and 199 (which were scrapped in 1929). The other, under B-892 (serials #26880-26881), was shipped in 1903 as GNRC 200 and 201. However, a Brooks builder's photograph exists of GNRC 70 with a serial #26866 – i.e. the number assigned to PM 199. The trade press in 1902 reported GNRC negotiations for two *passenger* 4-6-0s,

but a year later noted the delivery of two *freight* 4-6-0s. Historians surmise the GNRC originally placed two orders, likely to be GNRC 69-72, and then cancelled the first one (B-891), possibly because the company had been overly optimistic about needing both passenger and freight Ten Wheelers. It then appears that B-891 was diverted to fill the Pere Marquette order. By the time the second pair under B-892 was delivered, the CaNoR had commenced a new series of numbers in the 200-series for its 4-6-0s. Instead of continuing the GNRC series by using 71 and 72 (or reusing the cancelled numbers 69 and 70), they were delivered lettered and numbered GNRC 200 and 201. Both these 4-6-0s retained their (CaNoR) GNRC road numbers after the formal amalgamation into the Canadian Northern Quebec Railway.

CNR 1263-1267   4-6-0 TEN WHEEL TYPE   H-6-a

Cylinder	Gear	Driv.	Pressure	Boiler	T.E.	Haulage	Steam	Stkr.	Drivers/Eng./Total	Water	Coal	Length	Notes
			Specifications				Appliances		Weights			Length	Notes
20x26"	S	63"	200#	BEL	28060	28%	sat		130/160/258000	5000 gals	10 tons	- '	[1907]
20x26"	S	63"	200#	BEL	28060	28%	sat pv		130/160/268100	4500 gals	10 tons	- '	[1909-1913]
22x26"	S	63"	180#	BEL	30560	30>28%	H-C		126/157/274500	5000 gals	10 tons	61-1'	[CNR]
20x26"	S	63"	200#	BEL	28060	28%	H-C		126/157/300300	5800 gals	10 tons	65-1'	[B-26-a tender]

Brooks Locomotive Works – ALCO   1904   (B-993) $14,840   (5) Acquired by CNR 9-01-1919

	Serial	Delivered	New as	1-1912	Superheated	22x26"	Mods	Tenders	Disposition
				H-6-A				from	
1263	30184	11-15-04	**CaNoR 202**	CaNoR 1263	7-18 PK			B-26-a	Sc 8-28-44 PU
1264	30185	11-15-04	**CaNoR 203**	CaNoR 1264	8-18 PK			B-26-a	Sc 11-13-43 PU
1265	30186	11-15-04	**CaNoR 204**	CaNoR 1265	11-18 PK	11-18 PK		B-26-a	Sc 8-31-32 PU
1266	30187	11-10-04	**CaNoR 205**	CaNoR 1266	7-19 PK		w	B-26-a	Sc 5-10-45 PU
1267	30188	11-10-04	**CaNoR 206***	CaNoR 1267	12-15 PK			B-26-a	Sc 11-13-43 PU

CNR 1263-1267 were ordered for the **Canadian Northern Railway** on September 3rd 1904 and finished in October 1904. Alterations to both cylinders and boiler pressure were made on **1265** when it was superheated, although specifications reverted to the original when the cylinder bores were bushed after 1928. In the late 1930s, some in the group were retrofitted with tenders from retired B-26-a class 4-4-0s. The *CNR Mechanical Department Locomotive Diagrams* for **1261-1267**, from redrawn Issue L onward illustrated the class with a tender drawn for and with dimensions similar to B-26-a tenders.

The first pair built at Brooks for the GNRC had Belpaire fireboxes, inclined cylinders and inside steam pipes. The next five were built with the same specifications. Varnished and polished **CaNoR 206** (1267) at Dunkirk, New York in October 1904, had all of the usual CaNoR features.
[BROOKS WORKS PHOTO B-1287/GEORGE CARPENTER COLLECTION]

Equipped with a tender from a B-26-a 4-4-0, **1264**, at Nutana yard in Saskatoon on July 25th 1939, exhibited some of the usual CNR modifications, notably the altered cab, boiler tube pilot, and hanging running board steps. However, its appearance, along with all the others in the class, was significantly changed with the relocation of the sand dome and bell.
[EARL A. ELLIOTT/DON McQUEEN COLLECTION]

CNR 1268-1272   4-6-0 TEN WHEEL TYPE   H-6-b

Cylinder	Gear	Driv.	Pressure	Boiler	T.E.	Haulage	Steam	Stkr.	Drivers/Eng./Total	Water	Coal	Length	Notes
			Specifications				Appliances		Weights	Fuel Capacity		Length	Notes
20x26"	S	63"	200#	BEL	28060	28%	sat pv		130/160/268100	4500 gals	10 tons	- '	[1909-1913]
20x26"	S	63"	200#	BEL	28060	28%	H-C		126/157/274500	5000 gals	10 tons	61-1'	
20x26"	S	63"	200#	BEL	28060	28%	SCH		130/160/303100	5800 gals	10 tons	65-1'	[B-26-a tender]
22x26"	S	63"	180#	BEL	30560	30>28%	SCH		126/157/274500	5000 gals	10 tons	61-1'	

Locomotive & Machine Company of Montreal		1905	(Q-13)	$14,801							(5) Acquired by CNR 9-01-1919	
	Serial	Delivered	New as	1-1912	— Superheated —		22x26"	Tenders	Tenders	Disposition		
					H-C	SCH		from	to			
				H-6-B								
1268	30138	7-25-05	**CaNoR 207***	**CaNoR 1268**	11-17 PK					Sc 5-04-35 PU		
1269	30139	10-16-05	**CaNoR 208**	**CaNoR 1269**	9-17 PK					Sc 12-20-47 PU		
1270	30140	7-21-05	**CaNoR 209**	**CaNoR 1270**		11-13 PK		B-26-a	OCS	Sc 11-13-43 PU		
1271	30141	7-19-05	**CaNoR 210**	**CaNoR 1271**		4-13 PK		B-26-a		Sc 9-22-50 PU		
1272	30142	7-15-05	**CaNoR 211**	**CaNoR 1272**		4-15 PK	4-15 PK			Sc 9-29-32 PU		

CNR 1268-1277 were built under two orders for the **Canadian Northern Railway**: five in Q-13 and five in Q-15. Both orders were completed in December 1904, but shipped on the dates shown. All were built saturated with piston valves, an improved Belpaire firebox and wagon top boiler, and open cabs built with ash and pine. The original specifications for some were altered at the time they were superheated. By the early 1930s, they were converted back to 20x26" 200# by bushing the cylinder bores. The 30% haulage rating was then changed back to 28%, making it uniform with the rest of the H-6 class. In the late 1930s, some of this class were retrofitted with tenders from retired B-26-a class 4-4-0s. In May 1945, a wooden extension to the coal hopper was added to **1275**. The tender of **1270** was held for conversion to a water transport Car after the engine was scrapped.

In June 1956, authorization was given for **1274** to be renumbered, thus clearing the 1200-series for the new GR-12k SW1200RS diesel road switchers scheduled for delivery in 1957.

These two photographs of 1274 and 1270 (see next page) with that of 1276 in Vol. I, p. 101, taken throughout a decade, reflect a myriad of changes made to the ten locomotives in the H-6-b class. **CNR 1274**, at the Great Northern Railway coal plant in Vancouver during 1927, had been superheated with outside steam pipes but still retained the slatted pilot and tender lettering. If the person standing on the footboard was Cyril Littlebury, he either had a companion make the exposure or used his camera's delayed shutter release.
[CYRIL R. LITTLEBURY PHOTO?/
H.L. GOLDSMITH /GEORGE CARPENTER COLLECTION]

In 1961, **1521** was sold to Andrew MacLean of Don Mills for display at **Gravenhurst**, Ontario, but it became part of the proposed **Ontario Science Centre** project in 1965. When that collection was dispersed, it was purchased by R. Bennett of Croswell, Michigan on November 14th 1969 and shipped to him in March 1970. In 1989, it was acquired by the **Upper Clemens Theme Park** to become part of the display at Clementsport, Nova Scotia.

Built with a Belpaire firebox, 1270, at Saskatoon's Nutana yard on September 14th 1936, still had the as-built inside steam pipes and sloped cylinder block, a headlight mounted in front of the stack, but its running board configuration had been straightened (see 1276) and hanging ladders added. Its earlier tender had been replaced with one from a retired B-26-a 4-4-0.
[EARL A. ELLIOTT PHOTO/F.J. BECHTEL/DON McQUEEN COLLECTION]

CNR 1273-1277 — 4-6-0 TEN WHEEL TYPE — H-6-b

Cylinder	Gear	Driv.	Pressure	Boiler	T.E.	Haulage	Steam	Stkr.	Drivers/Eng./Total	Water	Coal	Length	Notes
			Specifications				Appliances		Weights	Fuel Capacity			
20x26"	S	63"	200#	BEL	28060	28%	sat pv		130/160/268100	4500 gals	10 tons	- '	[1909-1913]
20x26"	S	63"	200#	BEL	28060	28%	H-C		126/157/274500	5000 gals	10 tons	61-1'	
22x26"	S	63"	180#	BEL	30560	30%	SCH		126/157/274500	5000 gals	10 tons	61-1'	
20x26"	S	63"	200#	BEL	28060	28%	H-C/SCH		126/157/303100	5800 gals	10 tons	65-1'	[B-26-a tender] b

Locomotive & Machine Company of Montreal   1905   (Q-15) $14,801   (5) Acquired by CNR 9-01-1919

	Serial	Delivered	New as	1-1912	—— Superheated ——		22x26"	1-23-57	Mods	Disposition	To
				H-6-B	H-C	SCH		H-6-b			
1273	30562	7-28-05	CaNoR 212	CaNoR 1273		1-17 PK	b	1-17 PK		Sc 9-30-55 PU	
1274	30563	7-07-05	CaNoR 213	CaNoR 1274		2-14 PK	b	2-14 PK	CNR 1521/2	m	Dn 8-18-61 C	RAM
1275	30564	7-25-05	CaNoR 214	CaNoR 1275	12-17 AK	4-30 PU	b			Sc 1-11-55 PU	
1276	30565	6-20-05	CaNoR 215	CaNoR 1276	4-23 EH					Sc 7-26-35 MV	
1277	30566	7-18-05	CaNoR 216	CaNoR 1277	5-17 PK					Sc 10-31-32 PU	

CNR 1273-1277: see note under CNR 1268-1272 (page H-15).

CNR 1278-1287 — 4-6-0 TEN WHEEL TYPE — H-6-c

Cylinder	Gear	Driv.	Pressure	Boiler	T.E.	Haulage	Steam	Stkr.	Drivers/Eng./Total	Water	Coal	Length	Notes
			Specifications				Appliances		Weights	Fuel Capacity			
20x26"	S	63"	200#	EWT	28060	28%	sat		127/163/285000	5000 gals	10 tons	63-6½'	[Q-136/mod]
20x26"	S	63"	200#	EWT	28060	28%	sat pv		130/160/281000	5000 gals	10 tons	62-5'	[217-226: 1907]
22x26"	S	63"	180#	EWT	30560	30%	H-C/SCH		126/163/284800	5000 gals	10 tons	62-5'	□
20x26"	S	63"	200#	EWT	28060	28%	H-C/SCH		126/163/306600	5800 gals	10 tons	65-3'	[B-26-a tender] b
20x26"	S	63"	200#	EWT	28060	28%	H-C/SCH		126/163/277500	5000 gals	10 tons	65-1'	[H-5-a tender]

Locomotive & Machine Company of Montreal   1907   (Q-67) $19,890   (10) Acquired by CNR 9-01-1919

	Serial	Shipped	New as	1-1912	—— Superheated ——		Stl	( )-1957	Mods	Disposition	To
			—	H-6-C	SCH	H-C	cab	H-6-c			
1278	42652	5-13-07	CaNoR 217	CaNoR 1278	3-13 PK			CNR 1522/2 (4-30)	m	Sc 4-30-60 LM	
1279	42653	5-16-07	CaNoR 218	CaNoR 1279	5-30 PK	7-19 PK	□			So 5-11-42 PU	CR 74
1280	42654	5-20-07	CaNoR 219	CaNoR 1280		8-16 PK	□			Sc 7-31-54 PU	
1281	42655	5-20-07	CaNoR 220	CaNoR 1281	12-13 PK		b?		OCS	Sc 10-30-43 PU	
1282	42656	5-23-07	CaNoR 221	CaNoR 1282	6-13 PK		□			So 5-11-42 PU	CR 77
1283	42657	5-23-07	CaNoR 222	CaNoR 1283	5-14 PK/1-30		□			So 5-11-42 PU	CR 73
1284	42658	5-30-07	CaNoR 223*	CaNoR 1284	11-14 PK		□	CNR 1523/2 (1-16)	m	Sc 7-31-58 PU	
1285	42659	5-30-07	CaNoR 224	CaNoR 1285	8-13 PK		□			Sc 7-31-54 PU	
1286	42660	6-02-07	CaNoR 225	CaNoR 1286	6-13 PK/12-29		□		OCS	Sc 9-30-50 PU	
1287	42661	6-02-07	CaNoR 226	CaNoR 1287	12-16 PK		□	12-48 MP	m	Sc 7-14-55 LM	

H-6-b
H-6-c

The impression CNR changed only tender lettering and smokebox road number plates when it began to operate the CaNoR in 1919 may be oversimplified. **CaNoR 1304** was at Lucerne, British Columbia soon after the opening of that section of the transcontinental in August 1914. The most obvious alterations to **1302**, at an unidentified location during the late 1920s, were the addition of a triangular road number lamp and relettered cast number plate. Another change seen in other illustrations was CNR's use of a lower-case sub-class letter in the classification system, rather than one in the upper-case

as favoured by the CaNoR, but during the decade between the two photographs, other subtle changes were made by the CNR. The style of class lights and swing mechanism for the bell had been altered, while another modification was the location on the pilot of steam and air line hoses. The vertical stave pilots were early transitions to those using boiler tubes. However, the fundamental changes in appearance, as shown by the other H-6-c illustrations below, were to come during the next two decades.
[BOTH: AL PATERSON COLLECTION]

CNR 1278-1322 comprised three orders which totalled forty-five locomotives. Two of the purchases were made with the Locomotive & Machine Company of Montreal for ten (Q-67: **1278-1287**) ordered on October 31st 1906, and delivered in 1907, then twenty-five ordered on April 19th 1908, and shipped the same year (Q-79: **1288-1312**). The third lot (Q-136: **1313-1322**), ordered on March 31st 1910 and delivered later that year, was for ten from the Montreal Locomotive Works Ltd. (ALCO). CaNoR 227-251 (**1288-1312**) were built in the first two months of 1908, but shipped on the dates shown.

Schmidt superheating units were installed by CaNoR between March 1913 and January 1915, but beginning in November 1915, and until March 1924, only Hungerford & Cameron units were used. Eight of the latter were replaced with Schmidts. Three had their as-built Schmidt units replaced during 1929 and 1930. Original specifications were 20x26" 200# but, by March 1924, all save ten had been converted to 22x26" 180#. Between 1925 and 1931, twenty-seven had been reconverted to 20x26" 200# by bushing the cylinder bore and seven were retrofitted with 20x26-inch cylinder castings. See Figure HS (page H-21) for more detail. By 1930, those with a 30% haulage rating were changed to 28% to make them uniform with the rest of the H-6 class. Heavier frames (F) were installed into **1311** in July 1929 and **1317** in December 1931.

In the early 1940s, some were fitted with a tender from retired B-26-a 4-4-0s, and in the 1950s, **1300** acquired a tender from a retired H-5-a 4-6-0. A wooden extension to

the coal hopper was added to **1279** in January 1937 and to **1322** in June 1951.

During World War II the need for motive power in Australia became so acute, ten steam locomotives were purchased in North America by the **United States War Department** for the **Commonwealth Railways**. Of the total, eight were purchased from the CNR for $2759 each (becoming CR C^N 70-77), and two from the New York New Haven & Hartford Railroad, NH 846 and 820 (CR C^A 78-79). The eight from the H-6-c class were dismantled in Transcona and shipped on flat cars to the port of Vancouver. See Love & Matthews: *Canadian National in the West Vol. 4* p. 14 for more detail. The Canadian-built 4-6-0s remained in service until 1951 and 1952, the last one scrapped sometime after 1959. See Figure HC (page H-21). In 1954, **1319** was extensively damaged at Southampton, Ontario, in a derailment caused by a washout which occurred during Hurricane Hazel.

In June 1956, ten of the H-6-c class were assigned new numbers, to clear the 1200 and 1300-series for new SW1200RS diesel road switchers slated for delivery during 1957 and 1958. Only six were actually renumbered as the GR-12k, GR-12l and GR-12r classes arrived in 1957, early 1958, or in the last half of 1958. The four other steamers had been taken out-of-service before the arrival of the diesels, making renumbering unnecessary.

In 1957, **1315** was used for a time as a stationary boiler at Gravenhurst. The tender of scrapped **1304** was retained

(text continues on next page)

H-6-c

for an unspecified OCS assignment. The eventual service for the tender of **1289**, listed as held on December 31st 1955 for possible conversion to OCS assignments, remains uncertain. However, six others were slated for conversion to water transport Cars. In 1960, **1531** was donated to the **City of Barrie**, Ontario and was put on display in Centennial Park.

CNR 1278 to 1322 comprised the second-largest H class. **CaNoR 223** (1284) at Montreal in May 1907 carried typical company livery and appliances.
[MLW PHOTO Q-37/WES DENGATE COLLECTION]

CNR 1288-1312						4-6-0 TEN WHEEL TYPE						H-6-c
Specifications						Appliances		Weights		Fuel Capacity		Length	Notes
Cylinder	Gear	Driv.	Pressure	Boiler	T.E.	Haulage	Steam	Stkr.	Drivers/Eng./Total	Water	Coal		
20x26"	S	63"	200#		00	28%	sat pv		/ / 000	gals	tons	- '	[see CNR 1278-1287 (pg. H-16)]

			Locomotive & Machine Company of Montreal	1908	(Q-79)	$19,840					(25) Acquired by CNR 9-01-1919	
	Serial	Shipped	New as	1-1912	—— Superheated ——		Stl	Mods	( )-1957	Mods	Disposition	To
			—	H-6-C	SCH	H-C	cab		H-6-c			
1288	44775	7-20-08	CaNoR 227	CaNoR 1288		11-15 PK				OCS	Sc 10-31-50 PU	
1289	44776	7-20-08	CaNoR 228	CaNoR 1289	1-15 PK		□			OCS	Sc 11-21-47 PU	
1290	44777	7-14-08	CaNoR 229	CaNoR 1290	6-14 PK/2-30		□				So 5-11-42 PU	CR 76
1291	44778	7-14-08	CaNoR 230	CaNoR 1291		8-17 PK	□				Sc 12-19-35 PU	
1292	44779	7-14-08	CaNoR 231	CaNoR 1292	6-28 PK	6-17 PK	□				So 5-11-42 PU	CR 75
1293	44780	7-14-08	CaNoR 232	CaNoR 1293	8-29 PK	12-16 PK					So 5-11-42 PU	CR 70
1294	44781	7-13-08	CaNoR 233	CaNoR 1294		3-16 PK			(1524)/2		Sc 10-05-56 PU	
1295	44782	7-13-08	CaNoR 234	CaNoR 1295	2-15 PK		□			OCS	Sc 11-21-47 PU	
1296	44783	7-13-08	CaNoR 235	CaNoR 1296	5-13 PK		□				Sc 8-04-48 PU	
1297	44784	7-13-08	CaNoR 236	CaNoR 1297	5-14 PK		□				Sc 5-26-45 PU	
1298	44785	7-06-08	CaNoR 237	CaNoR 1298	10-13 PK		□				So 5-11-42 PU	CR 72
1299	44786	5-08-08	CaNoR 238	CaNoR 1299		9-17 PK	□				Sc 8-04-48 PU	
1300	44787	5-12-08	CaNoR 239	CaNoR 1300	10-29 PK	8-17 PK	□				Sc 8-21-55 PU	
1301	44788	7-06-08	CaNoR 240	CaNoR 1301		8-16 PK					Sc 7-05-54 PU	
1302	44789	6-30-08	CaNoR 241	CaNoR 1302		7-17 PK					Sc 5-19-45 PU	
1303	44790	6-18-08	CaNoR 242	CaNoR 1303	2-16 PK		□	9-34 MP m	**1525**/2 10-10	m	Sc 6-26-59 LM	
1304	44791	6-26-08	CaNoR 243	CaNoR 1304		8-16 PK	□			OCS	Sc 11-04-43 PU	
1305	44792	6-26-08	CaNoR 244	CaNoR 1305	8-14 PK		□				Sc 11-29-47 PU	
1306	44793	6-26-08	CaNoR 245	CaNoR 1306		8-19 PK	□			OCS	Sc 9-30-50 PU	
1307	44794	6-19-08	CaNoR 246	CaNoR 1307		9-18 PK	□	f	(1526)/2		Sc 6-14-57 PU	
1308	44795	6-19-08	CaNoR 247	CaNoR 1308	11-29 PK	12-16 PK	□				So 5-11-42 PU	CR 71
1309	44796	6-16-08	CaNoR 248	CaNoR 1309	11-29 PK	7-16 PK	□				Sc 7-31-54 PU	
1310	44797	5-14-08	CaNoR 249	CaNoR 1310	5-29 PK	12-15 PK	□				Sc 8-21-55 PU	
1311	44798	5-14-08	CaNoR 250	CaNoR 1311		8-16 PK	□	6-34 mF	**1527**/2 3-05-58	m	Sc 6-24-59 MP	
1312	44799	5-11-08	CaNoR 251	CaNoR 1312	3-13 PK/4-26 HQ		5-30 HQ m				Sc 2-11-55 PU	

CNR 1288-1312: see note under CNR 1278-1287 (page H-16).

Twenty years after being delivered from the builder, **1303**, at Fort Erie on July 18th 1927, was representative of CNR steam power before the introduction of the wafer.
[H.L. GOLDSMITH/ GEORGE CARPENTER COLLECTION]

Steel-cabbed **1312**, at the Stuart Street engine terminal in Hamilton on March 21st 1937, had undergone changes to the type of pilot and location of piping, while another modification was the relocation of the headlight.
[JOHN A. REHOR/ GEORGE CARPENTER COLLECTION]

H-6-c

CNR 1313-1322 — 4-6-0 TEN WHEEL TYPE — H-6-c

	Specifications						Appliances		Weights		Fuel Capacity		Length	Notes
Cylinder	Gear	Driv.	Pressure	Boiler	T.E.	Haulage	Steam	Stkr.	Drivers/Eng./Total		Water	Coal		
20x26"	S	63"	200#		00	28%	sat pv		/ / 000		gals	tons	63-6½'	[see CNR 1278-1287 (pg. H-16)]

Montreal Locomotive Works – ALCO   1910   (Q-136) $18,840   (10) Acquired by CNR 9-01-1919

	Serial	Shipped	New as	1-1912	── Superheated ──			Stl	Mods	Authorized 6-1956	Mods	Disposition	To
			—		H-6-C	H-C	SCH	cab		H-6-c			
1313	48309	7- -10	**CaNoR 252**	**CaNoR 1313**	2-24 EH □	SCH*		3-30 HQ				Sc 11-30-55 C	
1314	48310	-10	**CaNoR 253**	**CaNoR 1314**	2-23 EH			2-35 HQ (*11-39 HQ)				Sc 6-14-57 PU	
1315	48311	-10	**CaNoR 254**	**CaNoR 1315**	1-18 PK			10-32 HQ	fm	(1528)/2		Re 8-21-57 C	Sb
1316	48312	-10	**CaNoR 255**	**CaNoR 1316**	3-24 EH □			5-30 HQ		(1529)/2		Sc 10-05-54 AK	
1317	48313	-10	**CaNoR 256**	**CaNoR 1317**	6-23 EH □			9-37 MP	m F			Sc 4-13-54 PU	
1318	48314	-10	**CaNoR 257**	**CaNoR 1318**	7-17 PK			3-36 MP	m			Sc 1-11-55 PU	
1319	48315	-10	**CaNoR 258**	**CaNoR 1319**	11-17 PK □			11-30 HQ			Wr 10-15-54	Sc 11-14-55 LM	
1320	48316	-10	**CaNoR 259**	**CaNoR 1320**	11-16 PK □			4-34 HQ				Sc 8-30-55 LM	
1321	48317	-10	**CaNoR 260**	**CaNoR 1321**	10-15 PK □			5-30 HQ	mz	1530/2 10-10-56	mz	Sc 4-21-60 LM	
1322	48318	8- -10	**CaNoR 261**	**CaNoR 1322**	6-14 PK □	SCH		10-29	mz	1531/2 3-17-58	mz	Dn11-01-60 C	**ToB**

CNR 1313-1322: see note under CNR 1278-1287 (page H-16).

Commonwealth Railways **CR 76** (1290), one of eight from the H-6-c class 4-6-0s sold in 1942, was about to pull a freight out of Port Pirie, South Australia during May 1945. Before emigrating, modifications included outside steam pipes and running board ladders, but not a relocation of the headlight. Australian modifications added a tool box to the pilot deck, a wooden extension to the coal bunker, and the removal of the bell from the boiler check valve.
[J.B. GOGGS PHOTO/ANDREW MERRILEES COLLECTION/ LIBRARY AND ARCHIVES CANADA PAC E8222207]

Within two years of being photographed at Winnipeg's Fort Rouge engine terminal in August 1947, 1301 would have its headlight and bell relocated, but would retain its hanging running board ladders and inside steam pipes.
[DAVE SHAW/WES DENGATE COLLECTION]

H-6-c

Additional changes continued to take place to the H-6-c class during the last decade of service. **CNR 1294**, at Fort Rouge, Winnipeg, had been shopped on March 11th 1952 with a centred headlight and bell on top of the smokebox, but not with metal cab numerals. It had previously acquired outside steam pipes and running board ladders.
[DON McQUEEN COLLECTION]

Renumbered 2nd **1530** (ex 1321), at Palmerston, was similarly adorned as 1294, but by May 12th 1958 had both raised cab numerals and a horizontally-mounted tender wafer applied. [DOUG E. CUMMINGS/DON MCQUEEN COLLECTION]

FIGURE HS
CNR Class H-6-c (1278-1322) Specification Changes 1913-1931

The H-6-c class was built between 1907 and 1910 with 20x26" 200# specifications. Beginning in March 1913, CaNoR altered the specifications of some to 22x26" 180# as the 4-6-0s were fitted with either Schmidt or Hungerford & Cameron superheater units. The rationale for these changes in specification has not come to light in any records. Beginning in the early 1920s, CNR bushed some cylinders back to 20" and adjusted the boiler pressure back to 200#.

Rd. No.	Superheated 20x26 200#	Superheated 22x26 180#	Bushed to 20x26 200#	Rd. No.	Superheated 20x26 200#	Superheated 22x26 180#	Bushed to 20x26 200#	Rd. No.	Superheated 20x26 200#	Superheated 22x26 180#	Bushed to 20x26 200#
1278	3-13			1293	12-16			1308		12-16	1-28
1279		7-19	9-26	1294	3-16			1309		7-16	10-26
1280		8-16	9-29	1295		2-15		1310		12-15	4-27
1281	12-13			1296		5-13	1-29	1311		8-16	
1282		6-13	2-27	1297		5-14	10-26	1312		3-13	4-26
1283		5-14	11-26	1298		10-13	5-28	1313		2-24	
1284		11-14	3-27	1299		9-17	5-28	1314	2-23		
1285		8-13	8-28	1300		8-17	11-25	1315	1-18		
1286		6-13	11-26	1301	8-16			1316		3-24	11-31
1287		12-16	9-27	1302	7-17			1317		6-23	12-31
1288	11-15		(12-29)	1303		2-16		1318	7-17		
1289		1-15	4-27	1304		8-16	7-27	1319		11-17	2-26
1290		6-14	1-28	1305		8-14		1320		11-16	
1291		8-17	11-27	1306		8-19	9-28	1321		10-15	
1292		6-17	4-28	1307		8-18	6-27	1322		6-14	

FIGURE HC
Service record of the US War Department 4-6-0s on Australia's Commonwealth Railways.

CNR	CR	Retired	Sold as Scrap	Total CR Mileage	CNR	CR	Retired	Sold as Scrap	Total CR Mileage
1279	74	5-1951	3-27-1959	88,383	1292	75	10-1951	7-04-1958	95,141
1282	77	12-1951	after 1959	137,472	1293	70	12-1951	after 1959	96,632
1283	73	3-1951	3-27-1959	96,777	1298	72	5-1951	3-27-1957	123,804
1290	76	6-1952	after 1959	101,727	1308	71	5-1951	3-27-1957	84,416

NYNH&H	CR	Retired	Sold as Scrap	Total CR Mileage	NYNH&H	CR	Retired	Sold as Scrap	Total CR Mileage
846	78	-1946	3-27-1957	54,319	820	79	-1950	3-27-1957	117,730

H-6-c

CNR 1323-1342 — 4-6-0 TEN WHEEL TYPE — H-6-d

Specifications							Appliances		Weights	Fuel Capacity		Length	Notes
Cylinder	Gear	Driv.	Pressure	Boiler	T.E.	Haulage	Steam	Stkr.	Drivers/Eng./Total	Water	Coal		
22x26"	W	63"	170#	EWT	28060	28%	COLE		136/172/295340	5000 gals	10 tons	63-6½'	[MLW]
22x26"	W	63"	180#	EWT	30560	30>28%	H-C, SCH		136/172/295350	5000 gals	10 tons	63-6½'	

Montreal Locomotive Works – ALCO 1911 (Q-163) $19,939   (20) Acquired by CNR 9-01-1919

	Serial	Shipped	New as	1-1912	New	Stl	Mods	Authorized 6-1956	Ren. on	Mods	Disposition	To
			—	H-6-D	Superheater	cab		H-6-d				
1323	49876	5- -11	CaNoR 262	CaNoR 1323		6-28 HQ					Sc 4-14-54 LM	
1324	49877	5- -11	CaNoR 263	CaNoR 1324		7-30 HQ	mt	1532 /2	8-24-56	mtz	Sc 3-07-60 LM	
1325	49878	5- -11	CaNoR 264	CaNoR 1325	S: 1-52 HQ	9-39 HQ	m?	1533 /2	11-02-56	m z	So 11-12-62 C	NHIR
1326	49879	5- -11	CaNoR 265	CaNoR 1326		11-29 HQ					Sc 11-30-54 LM	
1327	49880	5- -11	CaNoR 266	CaNoR 1327	S: 4-32 PK		edm?	1534 /2	2-26-58	moz	Sc 9-21-61 PU	
1328	49881	5- -11	CaNoR 267	CaNoR 1328	S:12-31 PK		edm z	1535 /2	2-23-58	oz	Sc 2-14-60 PU	
1329	49882	6- -11	CaNoR 268	CaNoR 1329	S: 1-32 PK		ed				Sc 9-30-54 PU	
1330	49883	5- -11	CaNoR 269	CaNoR 1330	S: 7-30 PK		edm z	1536 /2	5-31-58	m z	Sc 2-14-60 PU	
1331	49884	6- -11	CaNoR 270	CaNoR 1331	H: 6-24 PK; S:12-29 PK	F					Sc 5-26-45 PU	
1332	49885	6- -11	CaNoR 271	CaNoR 1332	S:11-31 PK		e	(1537)/2			Sc 9-27-56 PU	
1333	49886	6- -11	CaNoR 272	CaNoR 1333	S: 6-41 PK		d z	1538 /2	9-26-58	z	Sc 4-07-60 PU	
1334	49887	6- -11	CaNoR 273	CaNoR 1334	H: 3-16 PK; S: 8-37 PK	F		(1539)/2			Sc 5-21-58 PU	
1335	49888	6- -11	CaNoR 274	CaNoR 1335	S: 1-30 PK	F z		(1540)/2			Sc 5-07-59 PU	
1336	49889	6- -11	CaNoR 275	CaNoR 1336	H: 3-25 PU; S: 7-45 PK		ed z	1541 /2	4-24-57	m z	Sc 10-14-61 LM	
1337	49890	6- -11	CaNoR 276	CaNoR 1337		12-36 MP	m	(1542)/2			Sc 8-24-56 LM	
1338	49891	6- -11	CaNoR 277	CaNoR 1338		6-36 HQ	m z	1543 /2	by 5-57	m z	Sc 10-04-57 LM	
1339	49892	6- -11	CaNoR 278	CaNoR 1339	S: 5-52 AV	9-38 MP	mtz	(1544)/3			Sc 2-21-58 AK	
1340	49893	6- -11	CaNoR 279	CaNoR 1340	S:10-39 HQ	9-31 MP	m	(1544)/3			Sc 3-13-59 LM	
1341	49894	6- -11	CaNoR 280*	CaNoR 1341		3-36 MP	m				Sc 6-12-56 AK	
1342	49895	6- -11	CaNoR 281	CaNoR 1342		3-37 MP	t				Sc 11-22-54 LM	

CNR 1323-1342 were ordered on January 3rd 1911 by the **Canadian Northern Railway**. *CNR Mechanical Department Locomotive Diagrams* Q-163 build-date listings show four in 1910 and sixteen in 1911. They were the first CaNoR 4-6-0s built with Walschaert valve gear. On September 2nd 1912, **CaNoR 1327** drew the special train carrying Alberta's new Governor General, the Duke of Connaught, into Edmonton for the official opening of the province's Parliament Building.

In two instances, the Hungerford & Cameron superheaters which had replaced the original Cole units, were themselves replaced by Schmidts. Heavier frames (F) were installed into **1331** (4-27), **1334** (3-48) and **1335** (12-47 PK). A wooden extension to the coal hopper (e) was added to **1327** (10-26), **1328** (2-26), **1329** (10-26), **1330** (9-27), **1333** (6-41) and **1336** (12-33).

A progression of images on pages H-22 to H-25 spanning four decades illustrate changes made to the H-6-d class. Walschaert-geared **CaNoR 280** (1341), at Montreal in June 1911, was superheated but had yet to be fitted with a turbo-generator and electrical system. [MLW PHOTO Q-163/H.L. GOLDSMITH/GEORGE CARPENTER COLLECTION]

In 1956, fourteen of the H-6-d class were assigned new numbers, to clear the 1300-series for new GR-12r and GR-12u SW1200RS diesel road switchers scheduled for delivery 1958 and 1959. Of those assigned new numbers, six were retired before renumbering took place. When **1328** was renumbered to **1535**, the metal cab numerals were replaced with painted Roman style numbers, similar to those currently being applied to diesels.

(text continues on next page)

CNR 1533 (1325) was sold from the Allandale scrap line in 1962 as **New Hope & Ivyland Railway** 1533, through the auspices of **United Scale Models Incorporated** (Steam Trains Inc.), of Chester, Pennsylvania. The locomotive was stored in Wilmington, Delaware and moved a year or two later to the Reading Company's St. Clair shops for an overhaul. Completely rebuilt by RDG and NHIR mechanics, 1533 was back in steam on July 2nd 1966, making a 110-mile transfer run to New Hope, Pennsylvania. It spent the next eight years as the workhorse of the NHIR hauling freight, passenger and mixed trains. Taken out-of-service in December 1975, it was scheduled to be rebuilt, but to date is stored unserviceable behind the New Hope shops.

By 1938, **1332**, at Saskatoon's Nutana yard, had been fitted with full running board ladders, extended coal bunker and carried its electric headlight on top of the smokebox. [SIRMAN COLLECTION]

CNR **1333**, at Winnipeg–Fort Rouge on July 30th 1950, had all of the refinements made to most of the class – the headlight centred on the smokebox face, bell in front of the stack, and the forward windows along the side of the cab blanked. Yet to come for most of the class were raised cab numbers and horizontally-mounted tender wafers. [AL PATERSON COLLECTION]

Two views of the engineer's side of the H-6-d class illustrate additional alterations made throughout the years of service. **CaNoR 1334** (1334) was at an unidentified location after its 1912 relettering, renumbering and classification. Although the extended piston rods and wood stave pilot are still in place, its headlight had been converted to electricity, with the turbo-generator tucked in between the headlight mounting and stack. The linkage on the pilot was to allow the engineer to control a pilot-mounted flanger or snowplow blade. **CNR 1338**, on the

Quebec Railway Light and Power electrified lines at St. Joachim, Quebec about 1938, was the only one in the class never to be fitted with full running board ladders. Gone were the wooden pilot and extended piston rods, and the layout had been altered for the power reverse cylinder and rodding. The turbo-generator had long been relocated ahead of the cab, and although the bell mount had been moved behind the check valve by the time of this photograph, it was never moved forward to the top of the smokebox. [BOTH: AL PATERSON COLLECTION]

H-6-d

CNR 2nd 1534 (1327), on an extra drag of empties, west of the St. James Jct. Interlocking tower on May 4th 1958, was one of two 4-6-0s with decal Roman numbers, most likely intended for a freshly-shopped first generation diesel road switcher.
[H. ROBERT (BOB) CLARKE PHOTO/AL PATERSON COLLECTION]

CNR 1343-1346   4-6-0 TEN WHEEL TYPE   H-6-e

Specifications							Appliances		Weights	Fuel Capacity		Length	Notes
Cylinder	Gear	Driv.	Pressure	Boiler	T.E.	Haulage	Steam	Stkr.	Drivers/Eng./Total	Water	Coal		
19x26"	S	63"	180#	WT	22790	23%	sat		115/149/244000	4000 gals	tons	- '	[1913]
22x26"	S	63"	180#	WT	30560	30>28%	H-C		126/157/274500	4000 gals	10 tons	61-1'	[rblt]
20x26"	S	63"	200#	WT	28060	28%	H-C		115/149/270500	5000 gals	10 tons	62-5'	[1345]

Brooks Locomotive Works   1901   (B-786) $15,000   (4) Acquired by CNR 9-01-1919

	Serial	Shipped	New as	7-1906	1-1912	Superheated	20x26"	Tenders	Disposition
			—	—	H-6-E			from	
1343	3773	2- -01	GNRC 57	(CaNoR) CNQ 57	CaNoR 1343	2-18 PK	9-27		Sc 11-05-35 PU
1344	3774	2- -01	GNRC 58	(CaNoR) CNQ 58	CaNoR 1344	12-17 PK	8-28		Sc 5-31-36 PU
1345	3775	2- -01	GNRC 59*	(CaNoR) CNQ 59	CaNoR 1345	3-18 PK	10-26	H-6-c	Sc 9-09-44 PU
1346	3776	2- -01	GNRC 60	(CaNoR) CNQ 60	CaNoR 1346	10-18 PK	10-25		Sc 6-17-41 PU

H-6-d

H-6-e

Almost four decades separate the two photographs of 1345 (see next page). As GNRC 59 (1345), at Dunkirk, New York in February 1901, the Belpaire firebox and inside steam ports of the inclined cylinder casting were to remain relatively unaltered during a myriad of other alterations.
[BROOKS WORKS PHOTO B-287/ALCO HISTORIC PHOTOS]

CNR 1343-1346 were built for the **Great Northern Railway of Canada** by Brooks ("Dunkirk" on *CNR Mechanical Department Locomotive Diagrams*). They retained their GNRC road numbers after amalgamation into the **Canadian Northern Quebec Railway**. The as-built specifications (20x26" 200#) were altered to 22x26", with 180# with modified cylinder castings at the time of superheating, although the internal steam pipes were retained. However, all were bushed back to 20x26" with 200# between 1925 and 1928. During the late 1930s, **1345** was fitted with a tender from a retired H-6-c 4-6-0.

During those thirty-eight years (see previous page), **1345** had been shopped with the usual improvements to the pilot, pumps, and cab. Although the headlight location remained unchanged, the kerosene lamp had been replaced with one powered by electricity. The check valves had been moved from the side to the top of the boiler, under the bell mount. During the process, the locations of the bell and sand dome were exchanged. An H-6-c tender had replaced the original by the time this photograph was taken at Brandon, Manitoba in September 1938. [EARL A. ELLIOTT/DON McQUEEN COLLECTION]

CNR 1347-1351   4-6-0 TEN WHEEL TYPE   H-6-f

Cylinder	Gear	Driv.	Pressure	Boiler	T.E.	Haulage	Steam	Stkr.	Drivers/Eng./Total	Water	Coal	Length	Notes
			Specifications				Appliances		Weights	Fuel Capacity		Length	Notes
22x26"	W	63"	170#	EWT	28632	28%	SCH		132/176/303300	6000 gals	10 tons	- '	[1913]
22x26"	W	63"	180>190#	EWT	32259	32>28%	SCH		132/176/303300	6000 gals	10 tons	67-0'	

Baldwin Locomotive Works   1911 $18,240   (5) Acquired by CNR 9-01-1919

	Serial	Shipped	Ordered as	New as	1-1912	To CNR	190#	Leased	Mods	6-23-59	Mods	Disposition
	—		—	—	H-6-F			1942=56		H-6-f		
1347	36933	9- -11	**DRL&W 975**	**DW&P 975**	**DW&P 1347**	5-27	5-33	NAR	f z	**1546** /3	mz	Sc 4-30-60 PU
1348	36934	9- -11	**DRL&W 976**	**DW&P 976**	**DW&P 1348**	5-27	11-33			(1547)/2		Sc 7-17-59 LM
1349	36935	9- -11	**DRL&W 977**	**DW&P 977***	**DW&P 1349**	5-27	5-36		z	(1548)/2		Sc 10-18-57 PU
1350	36967	9- -11	**DRL&W 978**	**DW&P 978**	**DW&P 1350**	10-27	7-32		mz	(1549)/2		Sc 12-21-61 LM
1351	36968	9- -11	**DRL&W 979**	**DW&P 979**	**DW&P 1351**	10-27	5-36			(1550)/2		Sc 5-31-59 PU

DW&P 977 (1349), at Philadelphia in September 1911, was one of the five in a class of locomotives transferred to the CNR in 1919.

[BLW PHOTO 3685/H.L. BROADBELT/WES DENGATE COLLECTION]

CNR 1347-1351 were ordered by the **Duluth Rainy Lake & Winnipeg Railway** on May 23rd 1911, but were delivered with **Duluth Winnipeg & Pacific Railway** lettering. The haulage rating of 30% (even though the tractive effort was 32,259 pounds) was reduced to 28% in 1924 to have it uniform for all H-6 classes. Between 1932 and 1936, the boiler pressure of 180# was increased to 190#. In 1927, all five were sold to CaNoR assets, in order to allow CNR to transfer and operate them in Canada. Prices ranged from $12,184 to $13,589, including import duty and sales tax. **CNR 1347** was known to have been leased on different occasions to the **Northern Alberta Railways** between 1942 and 1956. In 1956, all the H-6-f class was assigned new numbers, to clear the 1300-series for the future GR-12u SW1200RS road switchers slated for delivery during 1959. Only **1347** was actually renumbered as the remainder were out-of-service before the diesels arrived.

Twenty-five years later, on July 20th 1936, DW&P 977 as **CNR 1349** was steaming away on the Port Arthur waterfront. The extended piston rods had been removed but not the extended valve stems. The style of smoke stack and stave pilot had been changed, and a coal bunker extension added to increase capacity. The cab end of the running board had been raised to the as-built level of the midsection and extended to the smokebox. Maintaining the height predicated the use of a fixed four-step pilot ladder rather than the two- or three-step hinged versions.
[C.A. BUTCHER/DON McQUEEN COLLECTION]

At Allandale, a few months after its last shopping on January 11th 1958, **1350** had retained its high running boards, but had undergone cylinder modification, as well as the usual headlight and bell relocation. Raised cab numerals and horizontally-mounted tender wafers were the last modifications made to the Ten Wheeler.
[AL PATERSON COLLECTION]

H-6-c

DW&P 1352-1353 — 4-6-0 TEN WHEEL TYPE — H-7-a

Cylinder	Gear	Driv.	Pressure	Boiler	T.E.	Haulage	Appliances Steam	Stkr.	Weights Drivers/Eng./Total	Fuel Capacity Water	Coal	Length	Notes
19x26"	S	62"	190#	EWT	24449	24%	sat		111/143/265000	5000 gals	tons	- '	[1909]
19x26"	S	63"	180#	EWT	23220	23%	H-C		111/143/263000	6000 gals	10 tons	59-5'	

Rogers Locomotive Works – ALCO	1906	(J-1751)	$13,663				(2) Acquired by CNR 9-01-1919	
	Serial	Shipped	New as	3-1909	1-1912	Superheated		Disposition
			—	—	H-7-A			
1352	41210	11- -06	(ML&C) DRL&W 100	DW&P 910	DW&P 1352	5-21 PK		Sc 4-27-27 TD
1353	41211	11- -06	(ML&C) DRL&W 101	DW&P 911	DW&P 1353	11-20 PK		Sc 4-27-27 TD

DW&P 1352 and 1353 were built for the **Minnesota Land and Construction Company's Duluth Rainy Lake & Winnipeg Railway Company** which, in 1909, had been absorbed by the **Duluth Winnipeg & Pacific Railway.** In 1925, both 1352 and 1353 were retired but were returned to service later the same year.

With the only 4-6-0s remaining on the DW&P removed from the roster nine years after CNR began operating the line, and with no known builder's photograph in existence, two company images will have to suffice. Both **DW&P 1352** and **DW&P 1353** were possibly at West Virginia, Minnesota about 1923, shortly before their haulage rating was changed to 23%. The upper-case sub-class letter still remained as well, suggesting neither had received major shopping since the formation of the CNR. At this stage of their service careers they were still saturated, had horizontal stave pilots, graphited smokeboxes, open steel-sheathed cabs and retained their tender lettering. Their passenger train assignments may have been the reason for the generous use of white paint for wheel and rod trim.
[BOTH: *CNR LOCOMOTIVE DATA CARD*]

CNR 1354-1384 4-6-0 TEN WHEEL TYPE H-6-g

		Specifications					Appliances		Weights	Fuel Capacity		Length	Notes
Cylinder	Gear	Driv.	Pressure	Boiler	T.E.	Haulage	Steam	Stkr.	Drivers/Eng./Total	Water	Coal		
22x26"	W	63"	170#	EWT	28632	28%	SCH		133/173/297000	5000 gals	10 tons	-'	[1913]
22x26"	W	63"	180#	EWT	30560	30>28%	SCH		133/173/297000	5000 gals	10 tons	63-6½'	

Montreal Locomotive Works – ALCO 1912 (Q-188) $19,047   (31) Acquired by CNR 9-01-1919

	Serial	Shipped	New as H-6-G	Steel Cab	Leased 1942=56	To oil	Mods	To H-6-g	Mods/Tender to	Disposition	To
1354	50778	3- -12	CaNoR 1354	7-27 GP			mtz	1551 /2	10-31-56 m z	So 9-21-61 C	STM
1355	50779	3- -12	CaNoR 1355	6-29 HQ			edm	(1552)/2		Sc 11-21-57 LM	
1356	50780	3- -12	CaNoR 1356	2-39 HQ			m			Sc 9-17-54 LM	
1357	50781	3- -12	CaNoR 1357	8-35 HQ		7-58 MP x1359	m z	1553 /2	6-23-59 m	Sc 3-14-60 PU	
1358	50782	3- -12	CaNoR 1358	3-35 MP			edm			Sc 10-07-55 LM	
1359	50783	3- -12	CaNoR 1359	6-26 HQ	10-53 PK		edm	(1554)/2		Sc 7-14-58 PU	
1360	50784	3- -12	CaNoR 1360				m	(1555)/2		Sc 9-27-57 LM	
1361	50785	3- -12	CaNoR 1361	2-35 HQ			F m?			Sc 1-28-55 LM	
1362	50786	3- -12	CaNoR 1362	1-32 HQ		5-58 MP x1376	mtz	(1556)/2		Sc 3-25-60 PU	
1363	50787	3- -12	CaNoR 1363	7-35 HQ			edmt		OCS	Sc 10-14-54 LM	
1364	50788	3- -12	CaNoR 1364	12-35 MP			edmt	(1557)/2		Sc 11-08-57 LM	
1365	50789	3- -12	CaNoR 1365				edm?	(1558)/2		Sc 10-05-56 LM	
1366	50790	3- -12	CaNoR 1366	3-37 MP			edmt		OCS	Sc 10-14-54 LM	
1367	50791	3- -12	CaNoR 1367		10-53 PK		edm?	(1559)/2		Sc 11-14-58 PU	
1368	50792	3- -12	CaNoR 1368	10-36 MP			edm?t			Sc 8-15-55 LM	
1369	50793	3- -12	CaNoR 1369	12-36 MP			m			Sc 4-27-56 LM	
1370	50794	3- -12	CaNoR 1370				dmtz	1560 /2	11-17-56 mtz	Sc 3-14-60 LM	
1371	50795	3- -12	CaNoR 1371		NAR	8-58 MP x1382	ed z	(1561)/2		Ss 6-14-61 W	IPSCO
1372	50796	4- -12	CaNoR 1372				edm z	(1562)/2		Sc 12-26-58 PU	
1373	50797	4- -12	CaNoR 1373			9-54 PK	fedm	(1563)/2		Ss 10-01-59 W	IPSCO
1374	50798	4- -12	CaNoR 1374				mt r	1564 /2	8-25-56 mt r	Sc 4-21-60 LM	
1375	50799	4- -12	CaNoR 1375	4-27 AK			F m	1565 /2	11-06-57 m z	Sc 3-31-60 LM	
1376	50800	4- -12	CaNoR 1376			4-54 PK	m?t	(1566)/2		Sc 7-31-58 PU	
1377	50801	4- -12	CaNoR 1377			9-50 PK	m z	(1567)/2		Ss 10-01-59 W	IPSCO
1378	50802	4- -12	CaNoR 1378		NAR	9-55 PU	ed z	(1568)/2		Sc 10-06-58 PU	
1379	50803	4- -12	CaNoR 1379	12-43 MP			m			Sc 4-13-55 LM	
1380	50804	4- -12	CaNoR 1380			5-54 PK	F m?	(1569)/2		Sc 9-13-56 PU	
1381	50805	4- -12	CaNoR 1381				m	(1570)/2		Sc 7-31-58 PU	
1382	50806	4- -12	CaNoR 1382		NAR	2-54 PK	e	(1571)/2		Sc 11-14-58 PU	
1383	50807	4- -12	CaNoR 1383				ed z	(1572)/2		Sc 10-07-61 LM	
1384	50808	4- -12	CaNoR 1384			9-54 PK	fedm	(1573)/2		Ss 10-31-61 W	IPSCO

CNR 1354-1409 were built as **Canadian Northern Railway** 4-6-0s. The first thirty-one had been ordered on October 27th 1911 and the remaining twenty-five (1385-1409) on September 23rd 1912. The first thirty-one (1354-1384) were ordered with flangeless main drivers. The original haulage rating was reduced to 28% by

(text continues on next page)

Built under two orders from the same builder, the H-6-g class was to become the most numerous of all H classes. CaNoR 1381, one of thirty-one built in 1912, was apparently fresh from a shopping at Fort Rouge, judging from the reflection on the varnished cab and tender. Although the photograph was likely taken during World War I, the blind main drivers and as-built style of electric headlight remained unaltered, even though one set of spoked pilot wheels had already been changed.
[CANOR PHOTO/H.L. GOLDSMITH/GEORGE CARPENTER COLLECTION]

1930 to establish uniformity in all H-6 classes. During the 1940s, steel cabs were applied to those still in service.

Heavier steel frames (F) were fitted to eight of the group between 1924 and 1944. They included **1361** (12-40), **1375** (12-24), **1380** (5-31), **1386** (12-40), **1393** (1-41), **1395** (7-41), **1396** (1-37) and **1403** (12-44).

Between 1932 and 1952, another twenty-two were fitted with a wooden extension to the coal hopper (e). They included **1355** (6-52), **1358** (8-52), **1359** (6-37), **1363** (1-51), **1364** (7-52), **1365** (8-39), **1366** (4-52), **1367** (7-37), **1368** (11-52), **1370** (3-32), **1371** (6-32), **1372** (12-35), **1373** (11-32), **1378** (2-33), **1381** (9-34), **1383** (1-35), **1384** (3-32), **1389** (8-36), **1392** (8-39), **1397** (8-51), **1404** (8-39), **1409** (10-37).

Between 1942 and 1960, four were known to have been leased one time or another to the **Northern Alberta Railways**. Between 1950 and 1958, sixteen of the H-6-g class were converted to oil burners, four of which received tenders from retired members of the same class.

In June 1956, thirty-nine of the H-6-g class were authorized to be assigned new numbers, to clear the 1300-series for new SW1200RS diesel switchers in the future GR-12u and GR-12y classes, which were eventually delivered in 1959 and 1960. Only eight were actually renumbered, as the others were out-of-service before August 18th

Although **1373**, at Winnipeg's Fort Rouge shops on July 25th 1937, had yet to have headlight centring and bell relocated to the smokebox, it had acquired an extended coal bunker.
[AL PATERSON COLLECTION]

1960, the date of the arrival of diesel switchers 1396 and 1397, which marked the completion of CNR's dieselization program.

Between 1959 and 1961, four were sold as scrap to the **Interprovincial Steel Company**, although CNR records have listed **1371** and **1384** as initially being assigned Transcona for dismantling. Although sold on the date shown in the roster, **1377** was not turned over to the company until May 26th 1960. The tender of CNR **1399**, as **51534**, was used as an OCS auxiliary tender for Atlantic Region 50-ton crane **50143** [I-B 1902], assigned to Campbellton until retired for scrap in December 1972. The tenders of **1363** and **1366** were listed on December 31st 1955 as held for possible conversion to OCS assignments, but no other records list any details.

(text continues on next page)

CNR 1369, emerging from a smoke-hazed Turcot on April 13th 1946, had acquired a steel cab, flanged main drivers, running board ladders and centred headlight.
[H.L. GOLDSMITH/GEORGE CARPENTER COLLECTION]

CNR **1392**, retired on May 30th 1958, was donated a few weeks later to the **Edmonton Exhibition Association** for display at the Exhibition Grounds. In November 1970, it became part of the **Alberta Railway Museum** collection.

CNR **1395** and **1551** were sold to **F. Nelson Blount's Edaville Railroad Association** for display at **Steamtown** in Bellows Falls, Vermont. After the corporate move to Scranton, Pennsylvania, both were resold. **CNR 1551** went, in January 1986, to Jacob Jacobson who resold it to the **Ohio Central Railway** in 1988. In the same year, **1395** was purchased by the **Coopersville & Marne Railway** in Michigan.

CNR 1385-1409 — 4-6-0 TEN WHEEL TYPE — H-6-g

Cylinder	Gear	Driv.	Pressure	Boiler	T.E.	Haulage	Steam	Stkr.	Weights Drivers/Eng./Total	Water	Coal	Length	Notes
22x26"	W	63"	170#	EWT	28863	28%	SCH		133/173/297000	5000 gals	10 tons	- '	[1913]
22x26"	W	63"	180#	EWT	30560	30>28%	SCH		133/173/297000	5000 gals	10 tons	63-6½'	

Montreal Locomotive Works – ALCO 1913 (Q-213) $19,987   (25) Acquired by CNR 9-01-1919

	Serial	Shipped	New as H-6-G	Steel Cab	Leased 1942=56	To oil	Mods	To H-6-g	Mods/Tender	to	Disposition	To
1385	52642	5- -13	CaNoR 1385	6-28 HQ			m?				Sc 9-23-55 LM	
1386	52643	6- -13	CaNoR 1386	1-29 HQ			F m				Sc 11-30-54 LM	
1387	52644	5- -13	CaNoR 1387	10-26 HQ			m z	1574 /2 10-26-56	z		Sc 8-29-58 LM	
1388	52645	5- -13	CaNoR 1388	12-28 HQ			m				Sc 3-18-54 LM	
1389	52646	5- -13	CaNoR 1389	1-30 HQ	11-54 PK		edm z	(1575)/2			Sc 2-14-60 PU	
1390	52647	5- -13	CaNoR 1390	1-41 MP			mtz	1576 /2			Sc 3-21-60 LM	
1391	52648	5- -13	CaNoR 1391	4-29 HQ		6-58 MP x1367	m z	(1577)/2 8-30-56	mtz		Sc 5-07-60 PU	
1392	52649	5- -13	CaNoR 1392	10-29 HQ	NAR 10-54 PK		edwm	(1578)/2			Dn 6-17-58 W	Edm
1393	52650	6- -13	CaNoR 1393	6-29 HQ			F m	(1579)/2			Sc 3-20-59 LM	
1394	52651	6- -13	CaNoR 1394*	1-30 HQ			ef				Sc 10-18-56 LM	
1395	52590	4- -13	CaNoR 1395	2-20 HQ			F m z	(1580)/2	(r)		So 6-15-59 C	**STM**
1396	51591	4- -13	CaNoR 1396	7-40 HQ			F mtz	(1581)/2			Sc 1-23-59 MP	
1397	52592	4- -13	CaNoR 1397	7-37 MP			m z	(1582)/2			Sc 10-14-61 LM	
1398	51593	4- -13	CaNoR 1398	8-33 HQ			m				Sc 12-23-55 LM	
1399	52594	4- -13	CaNoR 1399	1-41 MP			m?		CN 51534		Sc 5-25-56 AK	
1400	52595	4- -13	CaNoR 1400	3-29 HQ			m				Sc 9-23-55 LM	
1401	52596	4- -13	CaNoR 1401	2-32 HQ			m z	(1583)/2			Sc 3-14-60 LM	
1402	52597	4- -13	CaNoR 1402	9-34 EH			m	(1584)/2			Sc 2-21-60 LM	
1403	52598	4- -13	CaNoR 1403				F m z				Sc 2-28-55 LM	
1404	52599	4- -13	CaNoR 1404	8-39 PK			edm?	(1585)/2			Sc 5-31-56 PU	
1405	52600	4- -13	CaNoR 1405	10-27 HQ			m	1586 /2 9-22-56	m		Sc 1-14-59 LM	
1406	52601	4- -13	CaNoR 1406	1-37 HQ			m z	(1587)/2			Sc 11-30-61 JB	
1407	52602	4- -13	CaNoR 1407	1-31 HQ			m?				Sc 7-14-55 LM	
1408	52603	5- -13	CaNoR 1408	2-36 HQ			m	(1588)/2			Sc 7-31-58 PU	
1409	52604	5- -13	CaNoR 1409	11-54 PK			ed	(1589)/2			Sc 6-01-59 PU	

The second batch of H-6-gs came in 1913. **CNR 1386**, at Montreal's Bonaventure Station in June 1941, was typical CNR in appearance for the class, with its centred headlight and the as-built running board hanging steps. Only three are known to not have their headlights centred (1365, 1372, 1378) and only one (1394) to have retained the hanging steps.
[AL PATERSON COLLECTION]

H-6-g

CNR 1385-1409: see note under CNR 1354-1384 (page H-29).

CNR 1389, ironically photographed beside the Fort Rouge, Winnipeg coal plant on August 12th 1959, was one of sixteen to be converted to oil during the 1950s. It was also one of many to receive metal cab numerals, and one of at least sixteen to have their tender wafers mounted horizontally. [AL PATERSON COLLECTION]

CNR 2nd 1564, at Palmerston on May 14th 1958, was one of eight renumbered, even though thirty-nine were assigned new road numbers. It was unique amongst all other 4-6-0s, in that it was shopped from Stratford on June 23rd 1956 (as 1374) with a circular maple leaf emblem on the tender. CNR 1395 also carried the same style of herald, but only during its cosmetic preparation in 1959 after its sale to Steamtown. [JOSEPH R. QUINN/GEORGE CARPENTER COLLECTION]

CNR 1410-1412 — 4-6-0 TEN WHEEL TYPE — H-8-a

Specifications						Appliances		Weights	Fuel Capacity		Length	Notes
Cylinder	Gear	Driv.	Pressure	Boiler	T.E.	Haulage	Steam	Stkr.	Drivers/Eng./Total	Water	Coal		
20x26"	S	63"	200#	EWT	28100		sat		122/155/279500	5000 gals	10 tons	60-4'	[CGR 1917]
20x26"	S	63"	200#	EWT	28100	28%	SCH		122/155/279500	5000 gals	10 tons	60-10'	

Canadian Locomotive Company Ltd.		1912	$19,500				(3) Acquired by CNR 9-01-1919	
	Serial	Shipped	Planned	New as	12-15-1915	Superheated		Disposition
			—	G4c 281%	T7-4c 140%	& EsC		
1410	1034	1-08-12	IRC 29	IRC 644	CGR 644	1-25 AK		Sc 4-30-37 AK
1411	1035	1-12-12	IRC 87	IRC 645	CGR 645	11-24 AK		Sc 6-08-36 AK
1412	1036	1-16-12	IRC 192	IRC 646*	CGR 646	3-26 AK		Sc 6-26-36 AK

CNR 1410-1412 were ordered in June 1911, in a broken series of road numbers, before the authorized renumbering of January 1912 went into effect on the **Intercolonial Railway of Canada**. They retained their IRC numbers during **Canadian Government Railways** operation. They were superheated using Schmidt units and Economy steam chests (EsC).

The disposition of the tenders from **1411** and **1412**, not scrapped with the engines at Moncton in 1936, is not recorded.

Of the forty-seven 4-6-0s with Intercolonial heritage, only three were assigned to CNR's H class. The remaining forty-three became members of the I class, and the last surviving 4-6-0 with 57-inch drivers went into the G class. One of the three of the future H class, at Kingston in January 1912, was saturated **IRC 646** (1412), with 63-inch drivers and inside steam pipes. [CLC PHOTO/DON McQUEEN COLLECTION]

As **1412**, it was likely in the scrap line at Moncton in 1936 at the end of its career. It had undergone few boiler-top changes, although the cab structure had been modified. However it had been superheated, utilizing Economy steam chests and outside steam pipes. [AL PATERSON COLLECTION]

CNR 1413-1418							4-6-0 TEN WHEEL TYPE							H-9-a
		Specifications					Appliances		Weights		Fuel Capacity		Length	Notes
Cylinder	Gear	Driv.	Pressure	Boiler	T.E.	Haulage	Steam	Stkr.	Drivers/Eng./Total		Water	Coal		
15½&26x28"	S	63"	200#	EWT	00		sat		/158/	000	gals	tons	- '	[CGW 4-6-0C 230-234]
20x28"	S	63"	200#	EWT	30225		sat		118/158/	000	gals	tons	- '	[CGW 4-6-0, 1901]
20x28"	S	62"	200#	EWT	30225		sat		118/163/	000	gals	tons	- '	[CGW 204 orig]
20x28"	S	63"	200#	EWT	30225		sat		/158/	000	gals	tons	- '	[CGW 204 mod. c.1909]
20x28"	S	63"	200#	EWT	30225		sat		118/158/267550		5000 gals	10 tons	64-6'	[CGR 1917]
20x28"	S	63"	200#	EWT	30225	30%	sat		118/163/272800		5000 gals	10 tons	65-0'	[CNR]

Baldwin Locomotive Works – Burnham, Williams & Company			1899	$17,785						(6) Acquired by CNR 9-01-1919	
	Serial	Shipped	New as	1901	3-1908	1917			To	Tender	Disposition
			—		—	T5-9	150%		1925	to	
1413	16640	4- -99	CGW 234	4-6-0	CGW 209	(GECo)	CGR 4541	6-08-17		CN 51570	Sc 10-31-27 AK
1414	16637	4- -99	CGW 231	4-6-0	CGW 206	(GECo)	CGR 4544	7-03-17			Sc 10- -25 AK
1415	16639	4- -99	CGW 233	4-6-0	CGW 208	(GECo)	CGR 4545	7-10-17	DPCo 1415		Sc 3-31-27 AK
1416	16638	4- -99	CGW 232*	4-6-0	CGW 207	(GECo)	CGR 4546	9-29-17			Sc 8- -25 AK
1417	16636	4- -99	CGW 230	4-6-0	CGW 205	(GECo)	CGR 4548	9-29-17			Sc 8-19-24 AK
1418	16580	3- -99	CGW 204				CGR 4549	10- -17		CN 50084	Sc 8-19-24 AK

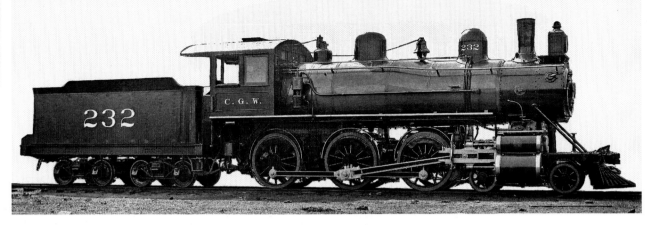

CNR 1413-1417 were built as Vauclain compounds for the **Chicago Great Western Railway** as 230-234, but converted simple in 1901. They were five of fourteen CGW locomotives sold to the **Canadian Government Railways** in 1917 through dealer **General Equipment Company**. In 1925, **1415** was either leased, sold, or leased with an option to purchase, to **Donnacona Paper Company**. It was returned by August 31st 1925 in exchange for CNR 2-6-0 E-7-a 847 and the 4-6-0 was put back on CNR records. The tender of **1413** was used as an OCS auxiliary tender **51578** until sometime after 1962 for the Atlantic Region Buildings and Bridges 35-ton crane 50125 [Ohio 1941].

CGW 232 (1416), at Philadelphia, was one of the Vauclain passenger compounds purchased in April 1899 by the Chicago Great Western Railroad.
[BLW PHOTO 1182/RAILROAD MUSEUM OF PENNSYLVANIA/ H.L. BROADBELT COLLECTION]

CNR 1418 was one of five built as simple 4-6-0s for the **Chicago Great Western Railway** as 200-204. See CNR 1543-1546 (page I-11). The tender of **1418** was used as the OCS auxiliary tender for the Atlantic Region 22-ton rail crane **50084** [O&S 1923] until it was scrapped in 1932. During this period, OCS tenders were given the same number as the cranes to which they were assigned.

At Winnipeg's Union Station sometime between June 1925 and October 1927, **1413** had been rebuilt simple by the CGW and purchased by the CGR for National Transcontinental Railway division assignments. Other than electrical upgrades and CNR front-end arrangement, only the running board alignment had been broken to accommodate the installation of an air brake reservoir and cooling piping.
[DETAIL: CNR PHOTO/AL PATERSON COLLECTION]

CNR 1419-1422						4-6-0 TEN WHEEL TYPE							H-9-b
Specifications						Appliances		Weights		Fuel Capacity		Length	Notes
Cylinder	Gear	Driv.	Pressure	Boiler	T.E.	Haulage	Steam	Stkr.	Drivers/Eng./Total	Water	Coal		
22&35x28"	S	63"	200#	EWT?	00		sat		/158/ 000	gals	tons	- '	[CGW 4-6-0C]
20x28"	S	63"	200#	EWT	30225	30%	sat		119/163/272800	5000 gals	10 tons	64-10'	

Rhode Island Locomotive Company		1900	$17,785								(4) Acquired by CNR 9-01-1919	
	Serial	Shipped	New as	1901	3-1908		1917			Tender	Disposition	
				—	—		T5-9	150%		to		
1419	3163	3- -00	**CGW 236**	4-6-0	**CGW 211**	(GECo)	**CGR 4550**	7-28-17			Sc 12- -25 EH	
1420	3166	3- -00	**CGW 239**	4-6-0	**CGW 214**	(GECo)	**CGR 4551**	7-28-17			Sc 10- -25 EH	
1421	3168	3- -00	**CGW 241**	4-6-0	**CGW 216**	(GECo)	**CGR 4552**	8- -17			Sc 10- -25 EH	
1422	3169	3- -00	**CGW 242**	4-6-0	**CGW 217**	(GECo)	**CGR 4553**	10- -17		OCS	Sc 9- -25 AK	

CNR 1419-1422 were in an order for ten Rhode Island compounds built as **Chicago Great Western Railway** 235-244 (#3162-3171), but converted simple in 1910. They were four of fourteen CGW locomotives sold to the **Canadian Government Railways** in 1917 through dealer **General Equipment Company**. The tender of **1422**, listed as marked (CGR) 4553, which may have been held for Atlantic Region OCS assignments, was scrapped in Moncton in May 1944.

Although not one of the four 4-6-0s to migrate to Canada seventeen years later, **CGW 235**, at Providence in February 1900, was shipped as a compound with inside steam pipes, extended piston rods and four-bar cross-head guides. At the time of the photograph, the windows had yet to be fitted.
[RHODE ISLAND WORKS PHOTO H-67/ALCO HISTORIC PHOTOS]

Nearing the end of its career at Pointe St. Charles shops, few major alterations appear to have been made to **1421** by either CGR or CNR before 1925. Although removed from the top of the smokebox, the headlight had been upgraded to electricity with the installation of the turbo-generator, mounted in the usual location on the top of the firebox, ahead of the cab.
[*CNR LOCOMOTIVE DATA CARD*]

CNR 1423 (first)   4-6-0 TEN WHEEL TYPE   first H-10-a

Specifications						Appliances		Weights		Fuel Capacity		Length	Notes
Cylinder	Gear	Driv.	Pressure	Boiler	T.E.	Haulage	Steam	Stkr.	Drivers/Eng./Total		Water	Coal		
18x26"	S	63"	170#		19000		sat		/ / 000		gals	tons	- '	

Grant Locomotive Works		18xx						(1) Acquired by CNR 9-01-1919
	Serial	Shipped	New as	??	??	3-1916		Disposition
			—	—	—	T 95%		
1423/1	???	-??	???	(???)	**OM&O 3**	**CGR 4503**		Sc 10-04-20 PU

CNR 1423 (first) was bought second-hand from an unidentified owner on an unrecorded date by the **O'Brien, McDougall & O'Gorman** (Construction) **Company**.

Although in the records as having been built by Grant Locomotive Works, no further builder's data has come to light.

CNR 1423-1442 (second/first) 4-6-0 TEN WHEEL TYPE second H-10-a

Cylinder	Gear	Driv.	Pressure	Boiler	T.E.	Haulage	Steam	Stkr.	Drivers/Eng./Total	Water	Coal	Length	Notes
			Specifications				Appliances		Weights	Fuel Capacity			
19x26"	S	63"	200#	EWT	25300		sat		126/167/304200	7000 gals	10 tons	65-1'	[GTP]
19x26"	S	63"	200#	EWT	25300	25%	SCH		126/171/317100	5800 gals	15 tons	64-1'	[coal bunker]
19x26"	S	63"	200#	EWT	25300	25%	SCH		126/171/317100	5800 gals	2600 gals	64-1'	[oil, orig] ■
19x26"	S	63"	200#	EWT	25300	25%	SCH		126/171/317100	5800 gals	3000 gals	64-1'	[oil: all others]

Montreal Locomotive Works – ALCO 1910 (Q-129) $17,500 (20) Operated and managed by CNR 7-12-1920; acquired 1-31-1923

	Serial	Shipped	New as	To oil	— Superheated —		Stl	Leased	Mods	6-56	Tender	Disposition
			A-1		EsC	W fwh	cab	1942-56		H-10-a	to	
1423/2	48032	5- -10	**GTP 600**	9-28 PK	9-20 PU		10-47		d w			Sc 12-02-55 PU
1424	48033	5- -10	GTP 601		3-18 RH				d			Sc 8-25-54 PU
1425	48034	5- -10	GTP 602	5-16 RH ■	8-21 PU		1-44 PK		d w?			Sc 1-14-55 PU
1426	48035	5- -10	GTP 603	10-15 PU	3-22 PU		?6-44 PU		d w			Sc 12-31-54 PU
1427	48036	5- -10	GTP 604	9-15 PU	3-19 RH		2-51 PK		d wm	(1590)/2		Sc 5-31-57 PU
1428	48037	5- -10	GTP 605	9-15 PU + 12-19 PU		11-23		d w			Sc 1-14-55 PU
				11-23 PU	11-23 PU							
1429	48038	5- -10	GTP 606	5-16 RH ■	6-22 PU				d w	(1591)/2		Sc 7-20-56 PU
1430	48039	5- -10	GTP 607	11-28 PK	6-22 PU		1-46 SH		d w			Sc 8-21-55 PU
1431	48040	5- -10	GTP 608		6-30 PU				d w		OCS	Sc 9-30-50 PU
1432	48041	5- -10	GTP 609	9-15 PU	11-21 PU				d wm			Sc 8-25-54 PU
1433	48042	5- -10	GTP 610	12-23 PU	7-18 RH				d w	(1592)/2		Sc 5-31-57 PU
1434	48043	5- -10	GTP 611	10-49 PK	8-19 RH			NAR	d w			Sc 6-04-54 PU
1435	48044	5- -10	GTP 612	5-16 RH ■	1-22 PU				d w?			Sc 1-14-55 PU
1436	48045	5- -10	GTP 613	4-16 PU	4-20 PU				d w			Sc 1-14-55 PU
1437	48046	5- -10	GTP 614	3-15 PU	3-22 PU				d w			Sc 9-24-54 PU
1438	48047	5- -10	GTP 615	6-22 PU	2-19 PU		9-47		d w	(1593)/2		Sc 7-20-56 PU
1439	48048	5- -10	GTP 616	2-22 PU	2-22 PU				d wf	(1594)/2		Sc 10-05-56 PU
1440	48049	5- -10	GTP 617	8-15 PU	1-19 RH				d w		Wr -52	Sc 12- -53 W
1441	48050	5- -10	GTP 618	10-23 PU	9-19 PU				d w		**CN 51534**	Sc 6-28-51 PU
1442	48051	5- -10	GTP 619	6-15 PU	5-18 SK				d w			Sc 8- -54 PU

+: Prince George shops (SJ) converted GTP 605 back to coal in 6-18.

CNR 2nd 1423, at Armstrong, British Columbia on July 21st 1945, was typical in appearance of the former GTP oil-burning 4-6-0s assigned to the BC District. The pocketing pole laying across the pilot would suggest switching assignments in need of long reaches. [JIM HOPE PHOTO/RAY MATTHEWS COLLECTION]

CNR 1423-1442 were ordered in January 1910 by **Grand Trunk Railway of Canada** for the **Grand Trunk Pacific Railway**. GTP converted thirteen from the order to oil between 1915 and 1922; CNR subsequently converted another five. GTP 605 was converted to oil twice, first in 1915 and again (as **1428**) in 1923, at the same time as a Worthington feedwater heater system was installed. The Worthington system was removed in June 1943. All superheater installations used Schmidt units and Economy steam chests with outside steam pipes. The first tenders used for the oil burners had a 2600 imperial gallon capac-

ity, but it eventually was increased to a uniform 3000 gallons. In May 1932, a wooden extension to the coal hopper was added to **1424**.

CNR 1426 had one of the most unique assignments of any CNR steamer. Fear of a Japanese invasion of British

(text continues on next page)

Columbia's Pacific coast during World War II resulted in the creation of an armoured train by the Canadian Army. Transcona shops completed the armouring of the rolling stock and fitted oil-fired 1426 with 8 mm armour plating over its wooden cab. The train and 1426 were finished on July 15th 1942, after which the Ten Wheeler hauled the special west, making the train's first operational run between Terrace and Prince Rupert on July 29th 1942. Initially it was planned both the pioneer oil (diesel) electric cab units of 1929, 9000 and 9001, would be armoured and re-engined in order to power the special train, but war priorities dictated only one could be done. By the time this was achieved in 1943, the armoured train was no longer needed, as the threat of invasion had failed to materialize. The train had made its final weekly run on September 17th 1943, after which orders were issued by Pacific Command Headquarters to temporarily suspend operation and put the train "in tallow". The train itself, less the locomotive, weapons and stores, remained at Terrace from September 25th 1943 until July 1944. The formal authorization to dismantle the train was issued by the Chief of General Staff on June 20th 1944. Once freed from armour train service, 1426 was re-fitted with a steel cab. So it was, despite the fact armour

In Saskatoon's Nutana yard in the early 1940s, 1434 was one of three in the Montreal-built lot to remain a coal burner and assigned to the Prairie District. It was converted to oil when assigned to the BC District in 1949.
[AL PATERSON COLLECTION]

plating was applied, and a new prime mover installed, that 9000 never did power the armoured train. For more about the train and the intended role of 9000 and 9001, see McQueen and Thomson: *Constructed In Kingston*.

Experimentation with dirigible (steerable) headlights (d) began in 1935, and by 1946 all had been fitted at least once with the device. See Appendix CF for specific details. **CNR 1434** was known to have been leased occasionally to the **Northern Alberta Railways** between 1942 and 1954. **CNR 1440** was sufficiently damaged in a collision with

(text continues on next page)

CNR 1427, possibly at Prince George in the BC District during the early 1950s, was one of at least six of the class known to have had their headlights centred and bells moved to the front of the smokebox. In addition to the wood stave pilot used in the district, it had acquired a steerable headlight and a set back three-step running board ladder to accommodate the extended smokebox. It was also one of the few to receive raised cab numerals.
[W.C. WHITTAKER/GEORGE CARPENTER COLLECTION]

CNR N-2-b 2496 at an unrecorded location in the Western Region in 1952 to warrant 1440's retirement in 1953. In June 1956, five of the group were assigned new numbers in anticipation of the 1400-series being used by diesel road switchers – but as the block of numbers was never used, the steamers were not renumbered.

The tender of 1431 was held for conversion to a water transport Car after the engine was scrapped. The tender of 1441 was used as an OCS auxiliary tender for Western

Oil-burning 1428, at Transcona in April 1933, had just emerged from the shops still with a Worthington feedwater system. The "lung", installed midway along the boiler on the fireman's side, was the only feedwater heater system applied to any CNR 4-6-0. The system was removed at Prince Rupert during June of 1943. [ROGER BOISVERT/AL PATERSON COLLECTIONS]

Region 40-ton pile driver 50122 [DH 1940]. The tender was then donated to the **Alberta (Pioneer) Railway Museum** at Edmonton.

CNR 1443-1452 — 4-6-0 TEN WHEEL TYPE — second H-10-a

Specifications							Appliances		Weights		Fuel Capacity		Length	Notes
Cylinder	Gear	Driv.	Pressure	Boiler	T.E.	Haulage	Steam	Stkr.	Drivers/Eng./Total		Water	Coal		
19x26"	S	63"	200#	EWT	25300		sat		126/171/304200		5800 gals	10 tons	65-1'	[GTP]
19x26"	S	63"	200#	EWT	25300	25%	SCH		126/171/317100		5800 gals	15 tons	54-8'	[hopper]
19x26"	S	63"	200#	EWT	25300	25%	SCH		126/171/314100		5800 gals	3000 gals	64-1'	[oil]

Canadian Locomotive Company		1910	$17,500						(10) Operated and managed by CNR 7-12-1920; acquired 1-31-1923				
	Serial	Shipped	New as	To oil	Coal	Superheated	Stl	Mods	6-56	Tender	Disposition	To
			A-1			EsC	cab		H-10-a	to		
1443	950	10-06-10	GTP 620	3-22 PU		3-22 PU	4-47 SH	d w		OCS	Sc 4- -53 PU	
1444	951	10-08-10	GTP 612	5-15 PU		6-22 PU		d w z	(1595)/2		Sc 9-30-61 PU	
1445	952	10-14-10	GTP 622			7-19 PU					Sc 12-06-47 PU	
1446	953	10-20-10	GTP 623	4-15 PU 5-22 SJ	4-21 PU	4-21 PU		d wfm	(1596)/2		Sc 6-21-56 PU	
1447	954	10-27-10	GTP 624*	4-15 PU		10-19 SK		d wf z	(1597)/2		Sc 9-30-61 PU	
1448	955	11-02-10	GTP 625	6-15 PU		12-22 PU		d wfm	(1598)/2		Sc 9-30-61 PU	
1449	956	11-05-10	GTP 626			11-17 PU			(1599)/2		Sc 12-06-47 PU	
1450	957	11-12-10	GTP 627			6-20 PU		f		OCS	Sc 11-21-47 PU	
1451	958	11-17-10	GTP 628	12-15 PU		10-20 PU	12-48 RH	d w m			Ss 10-01-59 W	IPSCO
1452	959	11-24-10	GTP 629			11-18 PU				OCS	Sc 12-20-47 PU	

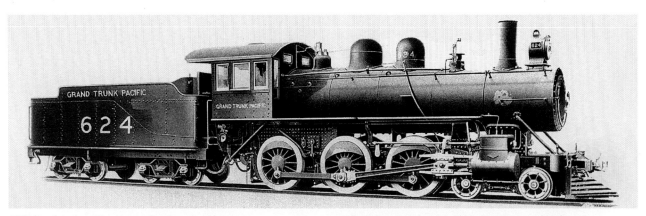

GTP 624 (1447), at Kingston in October 1910, was one in a similar order of 4-6-0s built the same year as those by Montreal. All thirty in the GTP A1 class were built simple and saturated with inside steam pipes, and with extended smokeboxes for improved heat combustion.
[CLC, MILN-BINGHAM LITHOGRAPH/DON McQUEEN COLLECTION]

CNR 1443-1452 were built for the **Grand Trunk Pacific Railway.** Between 1915 and 1922, six were converted to oil burners, four by GTP and two by CNR. As **GTP 605** (above), **623** underwent conversion to oil twice, first in 1915. It was then converted back to coal as **1445** when it was superheated in April 1921. Re-conversion to oil a second time took place a year later. Economy steam chests (EsC) with outside steam pipes were installed when the 4-6-0s were superheated. Installation of dirigible (steerable) headlights (d) began in 1940 for those assigned to the BC District. See Appendix CF for specific details. In June 1956, five were assigned new road numbers, to clear the 1400-series in anticipation for new diesel road numbers, but because it did not occur, none of the H-10s

As **1447**, at Armstrong, British Columbia on August 24th 1944, 624 had been superheated with outside steam pipes and Economy steam chests before becoming CNR stock, and later had been modified with hinged two-step running board ladders. The hooded headlight was a feature used exclusively in the BC District during World War II at a time when a Japanese invasion was anticipated.
[DICK GEORGE/WES DENGATE COLLECTION]

were actually renumbered.

All were scrapped by CNR between 1947-1961, except **1451**, which was sold as scrap to the **Interprovincial Steel Company** in 1959. The tenders of **1443, 1450** and **1452** were held for possible conversion to OCS assignments, but no records give any conversion information.

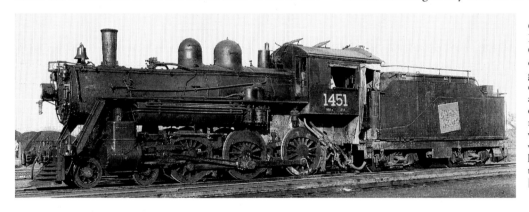

Oil-burning **1451**, at Fort Rouge, Winnipeg on October 10th 1957, was one of five in the CLC-built group which had been converted to oil five years after delivery. It was also one of three known to have raised cab numerals, but was not one of the two which eventually acquired horizontally-mounted tender wafers.
[ROBERT GUHR PHOTO/ AL PATERSON COLLECTION]

CNR 1453-1454 — 4-6-0 TEN WHEEL TYPE — H-11-a

Cylinder	Gear	Driv.	Pressure	Boiler	T.E.	Haulage	Steam	Stkr.	Drivers/Eng./Total	Water	Coal	Length	Notes
			Specifications				Appliances		Weights	Fuel Capacity		Length	Notes
13½&23x24"	S	62"	200#	EWT?	21380?		sat		97/126/216300	3200 gals	10 tons	58-0'	[4-6-0C]
19x24"	S	63"	180#	EWT	19000>21380		sat		97/132/220000	4000 gals	10 tons	58-9½'	[rblt 4-6-0]

Baldwin Locomotive Works – Burnham, Williams & Company			1897						(2) Acquired by CNR 7-16-1929	
	Serial	Shipped	New as	4-6-0	1909		1913		9-1927	Disposition
			SR-2		D4F		D4F			
1453	15473	9- -97	**CPR 483**/2	4-1906	**CPR 481**	1-09	**CPR 381**	9-13	**QM&S 300**	Sc 6-27-30 JD
1454	15478	9- -97	**CPR 488**/2	11-1905	**CPR 486**	2-09	**CPR 386**	11-13	**QM&S 301**	Sc 6-30-31 HQ

CNR **1453** and **1454** were two of thirteen Vauclain compound 4-6-0s built by Baldwin (#15470-15479 and #15521-15523) as **Canadian Pacific Railway I:480-492**/2 between September and October 1897. They became **CPR II:478-490** and **III:378-390** in the 1909 and 1912 renumbering respectively. The remainder of the lot was scrapped by CPR. For more details see Lavallée: *Canadian Pacific Steam Locomotives*. In 1927, two were sold to the **Quebec Montreal & Southern Railway,** just two years before that road's sale to the CNR. CNR records of December 31st 1930 show the boiler pressure of **1454** as 200 pounds.

Those familiar with Canadian Pacific motive power would recognize the lineage of CNR 1453 and 1454, despite the tender wafer and style of cab numbers. Although **CPR 482**, at Philadelphia in September 1897, was built as a balanced compound, early in its career it was rebuilt simple with inclined cylinders and inside steam pipes. [BLW PHOTO 975/RAILROAD MUSEUM OF PENNSYLVANIA, H.L. BROADBELT COLLECTION]

Even when sold, **QM&S 300** (1453), at CPR's Angus shops in September 1929, still retained its characteristic CPR smokebox front, tapered boiler, cab design and lettering style. [AL PATERSON COLLECTION]

Its mate, **QM&S 301**, as 1454 at Val Royal in July 1930, had some differences, notably with regard to location and style of headlight as well as type of pilot wheels, but was still recognizable as having been a CPR locomotive. [ROBERT R. BROWN PHOTO/ AL PATERSON COLLECTION]

1455-1497	Numbers not used
CNR 1498-1499	Numbers assigned but not used. See CNR **1111=1160** G-16-a class (page G-18)

CONTENTS

I CLASS: 1500-1628

THIS SECTION

I-1-a

I-1-b

I-2-a

I-2-b

I-3-a

I-3-b

I-3-c

I-4-a

I-4-b

I-5-a

I-5-b

I-6-a

I-6-b

I-7-a

I-7-b

I-8-a

CNR 1500-1628

I CLASS
4-6-0 TEN WHEEL TYPE

The "I" class was assigned road numbers 1500-1799 for Ten Wheelers with drivers larger than 63 inches. The class totalled 133 and came from two sources: eighty-six from **Grand Trunk Railway of Canada** (twenty-six from GTW and four which originally began service on the **Central Vermont Railway**), and forty-seven from **Canadian Government Railway** (forty-three of which had **Intercolonial Railway of Canada** origins). In 1919, the road number allotment assigned 1500-1546 to those of the CGR and in 1923, 1543-1628 were issued to the GTR Ten Wheelers. The overlap of numbers 1543-1546 was logical, as the CGR group was out-of-service before the GTR locomotive fleet was added to CNR stock. The "I" class was the only large CNR class in which resales or preservation did not occur. All dates are mm-dd-yr.

CNR 1500-1501 (first)   4-6-0 TEN WHEEL TYPE   I-1-a

Specifications							Appliances		Weights	Fuel Capacity		Length	Notes
Cylinder	Gear	Driv.	Pressure	Boiler	T.E.	Haulage	Steam	Stkr.	Drivers/Eng./Total	Water	Coal		
18x24"	S	70"	170#	BEL	16000	16%	sat		91/122/196500	3500 gals	7 tons	57-11'	[CGR, CNR]
18x24"	S	70"	200#	BEL	16000	16%	sat		91/122/207200	3500 gals	7 tons	58-5'	[1500]
18x24"	S	70"	170#	BEL	16000	16%	sat		91/122/207200	3900 gals	8 tons	58-5'	[1501]

Cooke Locomotive & Machine Company 1893						(2) Acquired by CNR 9-01-1919	
	Serial	Shipped	New as	1-04-1912	12-15-1915	Disposition	
			—	G7^A 162%	T2-7^A 80%		
1500/1	2293	9- -93	**IRC 171**/2*	**IRC 601**	**CGR 601**	Sc 3- -23 AK	
1501	2294	9- -93	**IRC 172**/2	**IRC 602**	**CGR 602**	Sc 3- -23 AK	

CNR **1500** and **1501** were built for the **Intercolonial Railway of Canada** with Belpaire fireboxes. The builder's record shows them delivered with 19½ x 24-inch cylinders. They retained their 1912 IRC road numbers during **Canadian Government Railways** ownership.

Incomplete **IRC 171** (1500) was rolled out onto the builder's transfer table at Paterson, New Jersey, about September in 1893 for its photograph. It and its mate were not only to become CNR's oldest passenger 4-6-0s, but ones with the shortest service life for the company. They were built saturated, with large diameter, evenly-spaced drivers, wooden cabs with three side windows, and short, low-capacity tenders. The Belpaire fireboxes installed in IRC 171 and 172 were the first of the design to be applied to a Canadian steam locomotive. [COOKE WORKS PHOTO C-109/GEORGE CARPENTER COLLECTION]

CNR 1502-1503 (first)   4-6-0 TEN WHEEL TYPE   I-1-b

Specifications							Appliances		Weights	Fuel Capacity		Length	Notes
Cylinder	Gear	Driv.	Pressure	Boiler	T.E.	Haulage	Steam	Stkr.	Drivers/Eng./Total	Water	Coal		
18x24"	S	70"	180#	BEL	17000	17%	sat		91/122/205400	4100 gals	8 tons	58-5'	[CGR, CNR]
18x24"	S	70"	180#	BEL	17000	17%	sat		91/122/205400	3000 gals	7 tons	58-5'	[1503]

Canadian Locomotive & Engine Company Ltd. (Dübs) 1894 $11,512						(2) Acquired by CNR 9-01-1919	
	Serial	Shipped	New as	1-04-1912	12-15-1915	Disposition	
			—	G7 174%	T2-7 85%		
1502	448	9-11-94	**IRC 173**/2	**IRC 603**	**CGR 603**	Sc 3- -23 AK	
1503	447	8-30-94	**IRC 125**/2	**IRC 604**	**CGR 604**	Sc 3- -25 AK	

CNR 1502 and 1503 were built for the **Intercolonial Railway of Canada** with Belpaire fireboxes. Tenders were placed in April 1893 for IRC 173 and 125. They retained their 1912 IRC road numbers during **Canadian Govern-**ment Railways ownership. *CNR Mechanical Department Locomotive Diagrams* have the CL&ECo serial numbers reversed for 1502-1503. **CNR 1503** was taken out-of-service in 1923.

CNR 1504-1506 (first)							4-6-0 TEN WHEEL TYPE							I-2-a
Specifications							Appliances		Weights		Fuel Capacity		Length	Notes
Cylinder	Gear	Driv.	Pressure	Boiler	T.E.	Haulage	Steam	Stkr.	Drivers/Eng./Total		Water	Coal		
19x24"	S	72"	180#	EWT	18400	18%	sat		97/135/219000		3500 gals	6 tons	60-11'	[CGR, CNR]
14&24x24"	S	72"	180#	EWT	18400?		sat		/136/	000	gals	tons	- '	[102 Cleveland]

Canadian Locomotive & Engine Company Ltd. (Dübs)			1898	$14,500						(3) Acquired by CNR 9-01-1919	
	Serial	Shipped	New as	1-04-1912	By 1914	12-15-1915		Tender to		Disposition	
			—	G5 187%		T3-5 90%					
1504	462	12-01-98	IRC 98/2	IRC 605		CGR 605			Rs 7-24	Sc 8-19-24 AK	
1505	463	12-26-98	IRC 102/2	IRC 606	4-6-0	CGR 606			Rs 7-24	Sc 12- -24 AK	
1506	461	11-15-98	IRC 167/2	IRC 607		CGR 607		CN 51550	Rs 7-24	Sc 12- -24 AK	

The three members of the I-2-a class were in CNR service for only five years. Two less-than-perfect images were taken for company records. **CNR 1504**, with the cover of its right-hand steam chest missing, was believed to be in Moncton about 1922.
[*CNR LOCOMOTIVE DATA CARD*]

CNR 1504-1506 were built for the **Intercolonial Railway of Canada**. IRC 102 (**CNR 1505**) was built with Cleveland cast iron cylinders as a test bed to compare their

(text continues on next page)

Out-of-service, likely at Moncton about 1924, **1505** had been built with experimental Cleveland cylinders. Arriving five years after IRC's first pair of passenger Ten Wheelers, the trio had larger, unevenly-spaced drivers and extended wagon top boilers.
[*CNR LOCOMOTIVE DATA CARD*]

efficiency to contemporary compounds. The elongated cylinders, moving the pistons on the uniflow principle, were the invention of William F. and Eugene W. Cleveland, of Reunthwaite, Manitoba. In the long run, however, the excessive weight of the cylinder saddle, as well as the advent of superheating, was likely a factor in IRC 102 being rebuilt simple prior to October 1914. (For more detail see page I-10 and Appendix BB in Vol. 1.)

CNR Mechanical Department Locomotive Diagrams show **CNR 1505** built in 1899. The 4-6-0s retained their 1912 IRC road numbers during **Canadian Government Railways** ownership. **CNR 1504-1506** were taken out-of-service in July 1923. The tender of the **1506**, renumbered to auxiliary tender **CN 51550** in 1924, was by 1962, assigned to the Chaleur Area Engineering Department based at Edmundston, New Brunswick. It was scrapped in July 1973.

CNR 1507 (first) — 4-6-0 TEN WHEEL TYPE — I-2-b

Cylinder	Gear	Driv.	Pressure	Boiler	T.E.	Haulage	Steam	Stkr.	Drivers/Eng./Total	Water	Coal	Length	Notes
14&24x24"	S	72"	180#	EWT	00		sat		/135/ 000	4300 gals	7 tons	60-11'	[Vauclain]
19x24"	S	72"	180#	EWT	18400		sat		97/135/228900	4300 gals	7 tons	60-11'	[CGR 1917]
19x24"	S	72"	180#	EWT	18400	18%	sat		97/135/228900	3500 gals	7 tons	60-11'	[CGR 1920]

Column headers: Specifications | Appliances | Weights | Fuel Capacity | Length | Notes

Baldwin Locomotive Works – Burnham, Williams & Company   1893 & 1895; rebuilt by ICR Moncton 1907   (1) Acquired by CNR 9-01-1919

	Serial	Shipped	New as	2-1898	To	1-04-1912	12-15-1915	Disposition
			4-6-0C	—	4-6-0	G5ᴬ 187%	T3-5ᴬ 90%	
1507	13320	3- -93	**BLW COLUMBUS**					Rb 8- -95 by BLW as
	14420	9- -95	**BLW 14420 ATLANTA***	**IRC 168/2***	-07 AK	**IRC 608**	**CGR 608**	Sc 7- -23 AK

CNR 1507, delivered as **Intercolonial Railway of Canada 168**, had been constructed in 1893 (under #13320) as a **Baldwin Locomotive Works** compound demonstrating the Vauclain system. Named *COLUMBUS*, it had been part of the BLW display at the **Columbian Exposition** in Chicago later the same year. The 4-6-0C was returned to the Baldwin plant and eventually rebuilt two years later with a new boiler under a new BLW serial. As **14420** *ATLANTA*, it was placed on display for the BLW exhibition at the **Cotton States and International Exposition** at Atlanta, Georgia in 1895.

After its return to Baldwin, it was imported into Canada during December 1897 for demonstration trials in the Montreal area and, by February 1898, was reported to have made successful runs between Moncton and Truro using much less coal and water than the other locomotives on the system. It was at this time the IRC finally purchased

To compare the efficiencies between the Cleveland and Baldwin compounds, IRC acquired a modified Vauclain demonstrator in February 1898. BLW 14420 *ATLANTA* (1507) was ready to leave Philadelphia in September 1895 for the trip to the Cotton States and International Exposition. [BLW PHOTO 853/RAILROAD MUSEUM OF PENNSYLVANIA, H.L. BROADBELT COLLECTION]

the compound and initially operated it as part of an evaluation with other locomotives having Cleveland cylinder systems. (See CNR 1505 on page I-3 and 1536 on page I-9.) *CNR Mechanical Department Locomotive Diagrams* list the 4-6-0 as built in 1907 – the date of its IRC rebuild. It is interesting to note Baldwin reused **1507**'s original boiler number (13320) for a 3'0" gauge 0-6-0ST shipped in March 1893 to the Southwest Virginia Improvement Company as number 11, *F.A. HILL.*

In February 1898, as **IRC 168** (1507), the 4-6-0 was ready to leave Baldwin for a third time. As a Vauclain passenger compound, the number and arrangement of high- and low-pressure cylinders were different from freight locomotives using the Vauclain system. For a comparison, see IRC 2-8-0 218 (1818) on page M-7. The angle of view in the builder's photograph was no doubt chosen to accentuate the power of the compound, especially with its over-sized cylinders and 72-inch drivers.
[BLW PHOTO/H.L. GOLDSMITH/GEORGE CARPENTER COLLECTION]

The illusion, however, evaporates with the broadside view of **1507** at an unidentified location about 1921.

[*CNR LOCOMOTIVE DATA CARD*]

CNR 1508 (first)						4-6-0 TEN WHEEL TYPE								I-2-b

		Specifications					Appliances		Weights	Fuel Capacity		Length	Notes
Cylinder	Gear	Driv.	Pressure	Boiler	T.E.	Haulage	Steam	Stkr.	Drivers/Eng./Total	Water	Coal		
19x24"	S	72"	180#	EWT	18400	18%	sat		97/135/228900	4300 gals	7 tons	60-11'	[CGR, CNR]

Baldwin Locomotive Works – Burnham, Williams & Company		1897	$10,000						(1) Acquired by CNR 9-01-1919
	Serial	Shipped	New as	1-04-1912	12-15-1915				Disposition
			—	G5^A 187%	T3-5^A 90%				
1508	15621	12- -97	**IRC 169**/2	**IRC 609**	**CGR 609**				Sc 8-19-24 AK

CNR 1508 was built as a simple 4-6-0 for the **Intercolonial Railway of Canada** and, as with 1507, was only reclassified while under **Canadian Government Railways** ownership. It was removed from service in July 1923.

CNR 1509-1513 (first)						4-6-0 TEN WHEEL TYPE								I-3-a

		Specifications					Appliances		Weights	Fuel Capacity		Length	Notes
Cylinder	Gear	Driv.	Pressure	Boiler	T.E.	Haulage	Steam	Stkr.	Drivers/Eng./Total	Water	Coal		
20x26"	S	72"	180#	EWT	22200		sat		114/145/240000	3000 gals	6 tons	60-7'	[orig]
20x26"	S	72"	180#	EWT	22200		sat		114/145/260040	3700 gals	8 tons	60-7'	[T4-4]
20x26"	S	72"	180#	EWT	22200		sat		114/145/266000	5000 gals	10 tons	60-3'	[T5-4]
20x26"	S	73"	180#	EWT	21783	21%	sat		114/145/266000	5000 gals	10 tons	60-3'	[CNR]

Canadian Locomotive & Engine Company Ltd. (Dübs)			1899-1900	$:various (see below)					(5) Acquired by CNR 9-01-1919
	Serial	Shipped	New as	orig.	1-04-1912	12-15-1915	1916-1919	1917 > 1919	Disposition
			—	cost	G4 222%	T4-4 110%			
1509	475	3-20-00	**IRC 72**/2	$13,500	**IRC 610**	**CGR 610**	CaNoR lease	T5-4 > T4-4	Sc 12- -25 EH
1510	476	4-05-00	**IRC 93**/2	$ 9,798	**IRC 611**	**CGR 611**	CaNoR lease		Sc 7- -25 AK
1511	472	12-03-99	**IRC 116**/2	$12,000	**IRC 612**	**CGR 612**			Sc 8- -25 AK
1512	474	12-06-99	**IRC 119**/2*	$12,000	**IRC 613**	**CGR 613**			Sc 7- -25 AK
1513	473	2-15-00	**IRC 166**/2	$12,000	**IRC 614**	**CGR 614**			Sc 12- -25 EH

I-3-a
I-3-b

CNR 1509-1513 were **Intercolonial Railway of Canada** 4-6-0s ordered in three lots by April 1899. By 1917, probably during the CaNoR lease, **Canadian Government Railways** 610 was reclassed to T5-4 when equipped with an oversize tender with a capacity of 5000 gallons of water, but, by 1919 it had been returned to class T4-4. All eventually were equipped with open steel cabs.

Within a year of the compounding experiments, IRC opted for simple, saturated 4-6-0s for its passenger trains, and continued to purchase similar locomotives for the next seven years. The future I-3-a class, built with larger cylinders and tenders, outweighed their predecessors by 19 tons. **CNR 1511**, likely at Moncton about 1925, with a horizontal-slatted pilot, a steel cab, and a two-piece window, was part of the order which followed the compounds.
[*CNR LOCOMOTIVE DATA CARD*]

CNR 1514-1517 (first)   4-6-0 TEN WHEEL TYPE   I-3-b

Cylinder	Gear	Driv.	Pressure	Boiler	T.E.	Haulage	Steam	Stkr.	Drivers/Eng./Total	Water	Coal	Length	Notes
			Specifications				Appliances		Weights	Fuel Capacity			
20x26"	S	72"	200#	EWT	24700		sat		114/145/269400	5000 gals	10 tons	61-3'	[CGR]
20x26"	S	73"	200#	EWT	24219	24%	sat		114/145/269400	5000 gals	10 tons	61-3'	[CNR]

Intercolonial Railway of Canada – Moncton   1903-1904   (unofficial serials) $14,576   (4) Acquired by CNR 9-01-1919

	Serial	Shipped	New as	1-04-1912	12-15-1915	1916-1919	Mods	Disposition
			—	G2 246%	T5-2 120%			
1514	(8)	6- -03	IRC 44/3	IRC 630	CGR 630	CaNoR lease		Sc 10-30-26 AK
1515	(9)	10- -03	IRC 46/3*	IRC 631	CGR 631		f	Sc 10-28-25 PU
1516	(10)	2- -04	IRC 48/3	IRC 632	CGR 632	CaNoR lease		Sc 12- -25 EH
1517	(11)	5- -04	IRC 49/3	IRC 633	CGR 633	CaNoR lease		Sc 10-28-25 PU

CNR 1514-1517 may have been ordered as early as March 1900 by the **Intercolonial Railway of Canada**.

All locomotives in the order eventually were equipped with open steel cabs.

CNR 1518-1521 (first) 4-6-0 TEN WHEEL TYPE I-3-b

			Specifications				Appliances		Weights	Fuel Capacity		Length	Notes
Cylinder	Gear	Driv.	Pressure	Boiler	T.E.	Haulage	Steam	Stkr.	Drivers/Eng./Total	Water	Coal		
20x26"	S	72"	200#	EWT	24700		sat		114/145/269400	5000 gals	10 tons	61-3'	
20x26"	S	73"	200#	EWT	24219	24%	sat		114/145/269400	5000 gals	10 tons	61-3'	[CNR]

Intercolonial Railway of Canada – Moncton		1907	(unofficial serials)	$14,207			(4) Acquired by CNR 9-01-1919	
	Serial	Shipped	New as	1-04-1912	12-15-1915	1916-1919		Disposition
			—	G2ᴬ 246%	T5-2ᴬ 120%			
1518	(12)	12- -07	IRC 16/4	IRC 634	CGR 634			Sc 10-15-28 EH
1519	(13)	12- -07	IRC 54/3	IRC 635	CGR 635			Sc 7-08-25 EH
1520/1	(14)	12?- -07	IRC 68/3	IRC 636	CGR 636	CaNoR lease		Sc 4- -25 EH
1521/1	(15)	12- -07	IRC 99/2	IRC 637	CGR 637			Sc 8-31-27 AK

CNR 1518-1521 were ordered in late 1905 by the **Inter-colonial Railway of Canada**. They were only reclassified while under **Canadian Government Railways** ownership. All eventually were equipped with open steel cabs.

IRC built its own passenger 4-6-0s to identical specifications in two lots, one in 1903-04 and another in 1907. Only scrap line views appear to exist of these derelict home-built Ten Wheelers in CNR livery. From the earlier lot, **1515** was photographed (on page I-6), likely at Transcona about 1925. **CNR 1519**, built as part of the later group, was likely at St. Malo about 1925. The inside steam pipes feeding the saturated cylinders remained unaltered during their two decades of service, although steel cabs replaced the originals. By the time of their retirement, 1515 had been modified with a footboard pilot, but still retained its IRC/CGR circular number plate. The semicircular handrail on the smokebox front of both 1515 and 1519 was so-mounted to permit maintenance of the kerosene headlights.
[BOTH: *CNR LOCOMOTIVE DATA CARD*]

CNR 1522-1529 (first) 4-6-0 TEN WHEEL TYPE I-3-c

			Specifications				Appliances		Weights	Fuel Capacity		Length	Notes
Cylinder	Gear	Driv.	Pressure	Boiler	T.E.	Haulage	Steam	Stkr.	Drivers/Eng./Total	Water	Coal		
20x26"	S	72"	200#	EWT	24600		sat		126/155/238700	3500 gals	7 tons	59-2'	[CGR]
20x26"	S	73"	200#	EWT	24219	24%	sat		126/155/238700	3500 gals	7 tons	58-10'	[CNR]

Manchester Locomotive Company		1901	$13,500				(8) Acquired by CNR 9-01-1919	
	Serial	Shipped	New as	1-04-1912	12-15-1915		Tender to	Disposition
			—	G1 246%	T6-1 120%			
1522/1	1746	3- -01	IRC 69/2	IRC 615	CGR 615			Sc 5- -25 AK
1523/1	1747	4-13-01	IRC 70/2	IRC 616	CGR 616			Sc 12- -25 EH
1524	1748	4-13-01	IRC 71/2	IRC 617	CGR 617			Sc 3- -25 AK
1525/1	1749	4-27-01	IRC 73/2	IRC 618	CGR 618		CN 50080	Sc 4- -25 AK
1526	1750	4- -01	IRC 74/2	IRC 619	CGR 619			Sc 7-08-25 EH
1527/1	1751	5- -01	IRC 76/4	IRC 620	CGR 620			Sc 4-30-27 AK
1528	1744	3-08-01	IRC 231	IRC 621	CGR 621			Sc 11-24-26 AK
1529	1745	3-28-01	IRC 232	IRC 622	CGR 622		CN 50820/1	Sc 4- -25 AK

CNR 1522-1529 were ordered during 1900 by the **Inter-colonial Railway of Canada**, and all except IRC 76 built in March 1901. IRC 76 was completed during April. After 1915, they all became **Canadian Government Railways** locomotives. All eventually were equipped with open steel cabs. The tender of the **1525**, renumbered to **CN 50080** in 1925 as an auxiliary water tender for Atlantic Region 15-ton crane 50080 [IBH 1920], was renumbered to **51125** in 1933. It was still in service in 1962, assigned to Moncton hump yard. In 1925 as well, the tender of

The eight 4-6-0s which came from Manchester in 1901 were approximately fourteen tons lighter than those built in Kingston or Moncton. Likely at Moncton about 1925, 1524 was retired with two types of pilot wheels, after-cooling piping on the outside of the air reservoir underneath the cab, and supports bolted to the tender frame in order to stabilize the rebuilt bunker and tank. [*CNR LOCOMOTIVE DATA CARD*]

the **1529** was renumbered to **CN 50820** as the auxiliary tender for the 35-ton Atlantic Region crane 50820 [Ohio 1924]. It was retired for scrap in 1932.

CNR 1529, at Moncton about 1925, was unadorned with piping under the cab, but retained its access steps to the sand dome and tender sides. [AL PATERSON COLLECTION]

CNR 1530-1534 (first) — 4-6-0 TEN WHEEL TYPE — I-4-a

	Specifications					Appliances		Weights	Fuel Capacity		Length	Notes
Cylinder	Gear	Driv.	Pressure	Boiler	T.E.	Haulage	Steam	Stkr.	Drivers/Eng./Total	Water	Coal		
20x26"	S	69"	200#	EWT	25600		sat		122/152/259000	4500 gals	9 tons	61-1'	[CGR]
20x26"	S	70"	200#	EWT	25257	25%	sat		122/152/259000	4500 gals	9 tons	60-8'	[CNR]

Canadian Locomotive Company	1904	$20,270					(5) Acquired by CNR 9-01-1919
	Serial	Shipped	New as	12-1904	1-04-1912	12-15-1915	Disposition
			—	—	G4^A 256%	T6-4^A 125%	
1530/1	621	7-04-04	**IRC 302**/1	**IRC** 1/4	**IRC 638**	**CGR 638**	Sc 12- -25 EH
1531/1	622	7-13-04	**IRC 303**/1	**IRC** 2/4	**IRC 639**	**CGR 639**	Sc 12- -25 EH
1532/1	620	6-29-04	**IRC 301**/1	**IRC** 6/4	**IRC 640**	**CGR 640**	Sc 3-31-30 AK
1533/1	624	7-29-04	**IRC 305**/1	**IRC** 43/3	**IRC 641**	**CGR 641**	Sc 12- -25 EH
1534/1	623	7-21-04	**IRC 304**/1*	**IRC** 50/3	**IRC 642**	**CGR 642**	Sc 10-29-25 PU

CNR 1530-1534 were built as **Intercolonial Railway of Canada** first 301-305 when shipped in June and July 1904. Within six months of their delivery they were renumbered to clear the 300-series for the renumbering of the new Consolidation Types of 1904 and the Pacific Types which were to arrive in 1905. See Appendix AA for the renumbering rationale. When the 4-6-0s were renumbered by the IRC in 1912, they retained the order of the IRC 1904 renumbering. The sequence remained unaltered when they were relettered for the **Canadian Government Railways**. All eventually were equipped with open steel cabs.

Following the four I-3-b class Ten Wheelers from Kingston was a batch of five 4-6-0s with smaller drivers than those of either the I-2 or I-3 classes. IRC 304 (1532), at Kingston near the end of June 1904, and the four others were initially numbered in a new series for a few months before being reassigned their second set of numbers. Three more renumberings would follow, but twenty-two years later, all but one had been removed from the CNR roster. [CLC, MILN-BINGHAM LITHOGRAPH FROM A HENDERSON PHOTOGRAPH/ DON McQUEEN COLLECTION]

CNR 1535 (first) — 4-6-0 TEN WHEEL TYPE — I-4-b

Cylinder	Gear	Driv.	Specifications Pressure	Boiler	T.E.	Haulage	Appliances Steam	Stkr.	Weights Drivers/Eng./Total	Fuel Capacity Water	Coal	Length	Notes
20x26"	S	69"	200#	EWT	25620		sat		122/152/259000	4500 gals	8 tons	58-8'	[CGR]
20x26"	S	70"	200#	EWT	25257	25%	sat		122/152/259000	4500 gals	9 tons	58-8'	[CNR]

Canadian Locomotive Company	1903	$15,270						(1) Acquired by CNR 9-01-1919
	Serial	Shipped	Ordered as	New as	1-04-1912	12-15-1915		Disposition
			—	—	G4ᴮ 256%	T6-4ᴮ 125%		
1535/1	579	4-29-03	AC&HB 28	**IRC 97**/2	**IRC 643**	**CGR 643**		Sc 5-31-27 AK

CNR 1535 had originally been ordered by the **Algoma Central & Hudson Bay Railway**. Construction was likely well in progress before the AC&HB cancelled the order,

and it was purchased by the **Intercolonial Railway of Canada**. It eventually was equipped with an open steel cab.

CNR 1536 (first) — 4-6-0 TEN WHEEL TYPE — I-5-a

Cylinder	Gear	Driv.	Specifications Pressure	Boiler	T.E.	Haulage	Appliances Steam	Stkr.	Weights Drivers/Eng./Total	Fuel Capacity Water	Coal	Length	Notes
20x26"	S	72"	200#	EWT	24600		sat		123/161/267300	4500 gals	9 tons	60-6'	[Cleveland cyl.]
21x26"	W	73"	180#	EWT	24550	24%	SCH		123/161/283000	5000 gals	10 tons	61-9'	[CNR, 1921]
21x26"	W	73"	180#	EWT	24031	24%	SCH		123/161/283000	5000 gals	10 tons	61-9'	[CNR, 1925]

Dickson Manufacturing Company	1901	$15,000						(1) Acquired by CNR 9-01-1919	
	Serial	Shipped	Received	New as	1-04-1912	12-15-1915	Superheated	Tender	Disposition	To
				—	G3ᴬ 246%	T8-3ᴬ 120%		to		
1536/1	1213	4-24-01	11-24-01	**IRC 233***	**IRC 623**	**CGR 623**	2-21 AK	OCS	Sc 6-26-36 AK	(BCCo)

CNR 1536, ordered in August 1900, was built as **Intercolonial Railway of Canada 233** with Cleveland simple cylinders. In September 1901, it attained 70 MPH in tests on the Delaware, Lackawanna & Western Railway out of Scranton, Pennsylvania, but some sources suggest the

Cleveland cylinders were removed before delivery to the IRC in November 1901, and IRC records of October 1914 did list it as having standard cylinders. Some of its earliest assignments were powering the *Maritime Express*

(text continues on next page)

between Truro and Moncton, as well as Halifax and Montreal. In 1912, as IRC 623, it was equipped with a larger tender than the G3 class. It was only reclassified while under **Canadian Government Railways** ownership. In February 1921, CNR extensively rebuilt **1536** with new cylinders, Walschaert valve gear, a Schmidt superheating unit and larger drivers with 73-inch diameters. The boiler of the scrapped 1536 was sold to the **Broughton Coal Company** of Sydney, Nova Scotia for $300. The disposition of **1536**'s tender, not scrapped with the engine, has not been discovered.

Three years after IRC 102 (1505) had been put into service with Cleveland cylinders, **IRC 233** (1536), at Scranton, Pennsylvania in April 1901, was added to the fleet. The massive inclined Cleveland cylinder system, appropriately plated, was fed by inside steam pipes. The wheel base of the engine truck had to be increased to accommodate the length of the cylinders. Unlike others built with standard cylinder housings (see IRC 62 on page I-11), 233 had only company initials painted on the cab rather than the full name. [DICKSON WORKS PHOTO D-4/GEORGE CARPENTER COLLECTION]

CNR 1537-1542 (first) — 4-6-0 TEN WHEEL TYPE — I-5-b

Cylinder	Gear	Driv.	Pressure	Boiler	T.E.	Haulage	Steam	Stkr.	Drivers/Eng./Total	Water	Coal	Length	Notes
20x26"	S	72"	200#	EWT	24600		sat		123/161/293600	4500 gals	9 tons	60-3'	[orig]
20x26"	S	72"	200#	EWT	24550	24%	sat		123/161/283000	4500 gals	9 tons	60-3'	[CNR, 1922]
21x26"	W	73"	180#	EWT	24300	24%	SCH		129/173/294600	5000 gals	10 tons	61-7'	[rblt]
21x26"	W	73"	180#	EWT	24031	24%	SCH		129/173/294600	5000 gals	10 tons	61-7'	[CNR, 1925]

Dickson Manufacturing Company   1901 $15,000 — (6) Acquired by CNR 9-01-1919

	Serial	Shipped	New as	3/4-1902	1-04-1912	12-15-1915	Superheated		Disposition
			—	—	G3 246%	T8-3 120%			
1537	1245	3-01-02	**IRC 234**		**IRC 624**	**CGR 624**			Sc 3- -25 AK
1538/1	1246	3-05-02	**IRC 235**		**IRC 625**	**CGR 625**	11-22 AK		Sc 6-08-36 AK
1539	1247	3-07-02	**IRC 236**		**IRC 626**	**CGR 626**	12-20 AK	Rs 12-31-31	Sc 12-20-32 AK
1540	1248	3-10-02	**IRC 61/3**	**IRC 237**	**IRC 627**	**CGR 627**			Sc 3- -25 AK
1541/1	1249	3-14-02	**IRC 62/3***	**IRC 238**	**IRC 628**	**CGR 628**	12-22 AK		Sc 6-08-36 AK
1542	1250	4-04-02	**IRC 63/3**	**IRC 239**	**IRC 629**	**CGR 629**	3-21 AK		Sc 6-26-36 AK

CNR 1537-1542 were built for the **Intercolonial Railway of Canada**, although IRC 61-63 were renumbered immediately after delivery to avoid conflict with second IRC 61-63 (Manchester 4-4-0s of 1875) which were still in service. IRC 239 was wrecked in 1904 at Fall River, Nova Scotia, near Windsor Jct., one of several accidents which resulted in crews nicknaming it the "hoodoo engine". After 1915, the 4-6-0s were transferred to **Canadian Government Railways** ownership. In July 1920, CGR 234 (**CNR**

1537) was in a wreck at Coldbrook, New Brunswick and was subsequently repaired by CNR at Moncton. When superheated, all were rebuilt with new cylinders, Walschaert valve gear, and fitted with 73-inch drivers. Specification data for the rebuilds is based upon the 5000-gallon tender tank, even though capacities within the I-5-b class varied.

The disposition of the tenders from **1538**, **1541** and **1542**, not scrapped with the engine at Moncton in 1936, is not known.

IRC 62 (1541) at Scranton, Pennsylvania on or about March 14th 1902, was renumbered by the railway within weeks of its arrival from the builder. Although undergoing some modifications, it was one of two in the order which retained its Stephenson valve gear and was never superheated. [DICKSON WORKS PHOTO D-22/GEORGE CARPENTER COLLECTION]

CNR 1538, at Moncton in May 1936, had been rebuilt with Walschaert valve gear and superheated with cast cylinder housings and outside steam pipes. [DON McQUEEN COLLECTION]

I-5-b

I-6-a

CNR 1543-1546 (first, second) 4-6-0 TEN WHEEL TYPE first I-6-a

			Specifications				Appliances		Weights	Fuel Capacity		Length	Notes
Cylinder	Gear	Driv.	Pressure	Boiler	T.E.	Haulage	Steam	Stkr.	Drivers/Eng./Total	Water	Coal		
20x28"	S	62"	200#	EWT	00		sat		/163/ 000	gals	tons	- '	[orig CGW]
20x28"	S	63"	200#	EWT	00		sat		/163/ 000	gals	tons	- '	[CGW reclassed, 1909]
20x28"	S	63"	200#	EWT	00	30%	sat		123/163/280630	5100 gals	9 tons	65-10'	[CNR 1920; I-6-a]
20x28"	S	68"	200#	EWT	28000	28%	sat		123/163/280630	5100 gals	10 tons	65-4'	[T7-9, H-2-c]

Baldwin Locomotive Works – Burnham, Williams & Company   1899 $17,785   (4) Acquired by CNR 9-01-1919

	Serial	Shipped	New as		1917		1920			Disposition
					T7-9 140%		H-2-c			
1543/1	16576	3- -99	CGW 200	(GECo)	CGR 4540	6-18-17	CNR 1203/1	4- -20		Sc 3- -25 AK
1544/1	16579	3- -99	CGW 203	(GECo)	CGR 4542	6-23-17	CNR 1204/1	4- -20		Sc 12- -25 EH
1545/1	16578	3- -99	CGW 202	(GECo)	CGR 4543	6-23-17	CNR 1205/1	3- -20		Sc 12- -25 EH
1546/1	16577	3- -99	CGW 201	(GECo)	CGR 4547	8- -17	CNR 1206/2	3- -20		Sc 12- -25 EH

CNR 1543-1546 (first, second) were built as **Chicago Great Western Railway** 200-204. They were five of fourteen CGW locomotives sold to **Canadian Government Railways** in 1917 through dealer **General Equipment** Company. CGW 204 was not included in any sale to a Canadian railway. CGW had converted 200-203 to 68-inch drivers in 1909 but in 1920, CNR reduced driver size to 63-inches and reclassed them from I-6-a to H-2-c.

Tucked in behind a pole and switch stand, **CNR 1st 1203**, photographed at Moncton possibly in 1925, had no class or haulage rating painted on the cab. The peeling smokebox and lack of a main driving rod suggest it had already been removed from the roster. The larger-than-usual spacing between the boiler and the drivers was created when the latter were reduced in size by five inches in 1920, thus altering its class designation from an "I" to "H". [H.L. GOLDSMITH/GEORGE CARPENTER COLLECTION]

FIGURE IC
A Summary of 4-6-0s built for the Grand Trunk Railway of Canada between 1891 and 1908

The backbone of GTR's premiere passenger locomotive fleet between 1898 and the introduction of the Pacific Types in 1910 was the A class 4-6-0. GTR either built, or had built, eighty-six 4-6-0s during a eleven-year span, and, in 1923, all eighty-six were taken into CNR stock. A summary of both the CNR G and I class locomotives follows, arranged in sequential order of GTR acquisition.

The CNR renumbering plan for the I-6 and I-7 classes was based upon the current GTR classification rather than the GTR road number. The numbering sequence began with the A8 class, followed by the A4, and then the A10. Allowances were made for builders' lots and saturated engines not yet superheated. GTR 331 and 332 (1554 and 1555) were placed with four other Schenectady-built 4-6-0s (GTR 304-307) for continuity, while GTR 335 (1581) and 340 (1573) were renumbered in sequence according to their GTR specifications, even though in 1923 both were still class A 4-6-0s.

GTR A8	CNR I-6-a	1543-1545	GTR A6	CNR I-8-a	1589-1628
GTR A4	CNR I-6-b	1547-1577	GTR A10	CNR I-7-a	1578-1588

Lot	Built			New as		1910	A	A1	A2	A3	A4	A4*	A6	A7	A8	A8*	A10	A10+	A10	CNR	To	
1	1891	BLW	2	C&GT	152-153	2340-2341		A1													1167	G-19-a
2	1898	BLW	4	GTR	992	300	A				A4	A4*			A8						1543/2	I-6-a
					993-995	301-303	A				A4	A4*									1547-1549	I-6-b
3		Schen	4	GTR	996-999	304-307	A				A4	A4*									1550-1553	I-6-b
4	1899	PSC	6	GTR	986	308	A								A8						1544/2	I-6-a
					987-990	309-312	A				A4	A4*									1556-1559	I-6-b
					991	313	A								A8						1545/2	I-6-a
5	1901	PSC	12	GTR	978-979	314-315	A				A4	A4*								16b	1560-1561	I-6-b
					980	316	A										A10	A10*			1578	I-7-a
					981-982	317-318	A				A4	A4*									1562-1563	I-6-b
					983	319	A											A10+			1583	I-7-b
					984-985	320-321	A				A4	A4*								16b	1564-1565	I-6-b
				CVR	220 GTR	322	A											A10+			1584	I-7-b
				CVR	221 GTR	323	A				A4	A4*									1566	I-6-b
				CVR	222 GTR	324	A				A4	A4*									1567	I-6-b
				CVR	223 GTR	325	A											A10+			1585	I-7-b
6	1902	PSC	5	GTR	973	326	A											A10+			1586	I-7-b
					974-977	327-330	A				A4	A4*									1568-1571	I-6-b
7	1904	ALCO-S	5	GTR	964-965	331-332	A				A4	A4*									1554-1555	I-6-b
					966-968	333-335	A										A10			I7a	1579-1581	I-7-a
8	1904	MLW	10	GTR	954	336	A											A10+	A10†		1587	I-7-b
					955	337	A					A4*								16b	1572	I-6-b
					956	338	A											A10+			1588	I-7-b
					957	339	A								A8	A8*					1546/2	I-6-a
					958-962	340-344	A				A4	A4*								16b	1573-1577	I-6-b
					963	345	A										A10				1582	I-7-a
9	1905	BLW	10	GTR	1352-1361	1640-1649			A2					A7							1168-1177	G-20-a
10	1906	MLW	10	GTR	1000-1009	400-409				A3			A6							I8a	1589-1598	I-8-a
11		ALCO-S	10	GTR	1010-1019	410-419				A3			A6								1599-1608	I-8-a
12	1907	PSC	10	GTR	1020-1029	420-429				A3			A6							I8a	1609-1618	I-8-a
13	1908	BLW	10	GTR	1030-1039	430-439				A3			A6								1619-1628	I-8-a
			86				A 46	A1 2	A2 10	A3 40	A4 31	A4* 29	A6 40	A7 10	A8 4	A8* 1	A10 4	A10+ 5	A10 2	CNR 7		

I-6-a

FIGURE IS
Specifications of Grand Trunk Railway of Canada's A class

Class	Total	1st use	— Basic Specifications —						TE		Weight		Lots	Illustrations	Built
		1-1898	20x26"	S	D	ins	73"	200#	25425	sat	308628		2	GTR 992 p.I-14	1898
A	46	11-1904	20x26"	S	D	ins	73"	200#	24219	sat	308628	Light A	2-6	GTR 986 p.I-15	1898
			20x26"	S	PV	ins	73"	200#	24219	sat	308628	Light A	6		1899
			20x26"	S	D	ins	73"	210#	24219	sat	312856	Heavy A	7	GTR 967 p.I-18	1902
			20x26"	S	D	ins	73"	225#	24219	sat	317770	Heavy A	8	GTR 959 p.I-22	1904
		6-1907	20x26"	S	D	ins	73"	210#	24219	sat	312856	Heavy A	8		1904
		2-1910	20x26"	S	D	ins	73"	200#	24219	sat	312856	Heavy A	7,8		1904
A1	2	11-1904	18x24"	S	D	ins	57"	160#	18553	sat	190000		1	GTR 1241 p.G-23	1891
A2	10	1-1905	14&24x26"	S	D	ins	56"	180#	23000	Vc	229000		9	GTR 1640 p.G-24	1905
		by-1910	19x26"	S	D	ins	56"	180#	26000	sat	298052		9		1905
A3	40	6-1906	19x26"	S	D	ins	73"	210#	21858	sat	286000		10,11	GTR 401 p.I-26	1906
		2-1910	19x26"	S	D	ins	73"	200#	21858	sat	300140		10,11	GTR 411 p.I-29	1906
		5-1907	19x26"	S	D	ins	73"	210#	21858	sat	300140		12		1907
		2-1910	19x26"	S	D	ins	73"	200#	21858	sat	300140		12		1907
		5-1908	19x26"	S	D	ins	73"	210#	21858	sat	288000		13		1908
		2-1910	19x26"	S	D	ins	73"	200#	21858	sat	288000		13		1908
A4	17	8-1914	22x26"	S	PV	out	73"	180#	26375	SCH	308628	Light A4	2-6		1898-1902
		7-1915	22x26"	S	PV	out	73"	180#	26375	SCH	312856	Heavy A4			1904
A4*	27	10-1917	21x26"	S	PV	out	73"	180#	24032	SCH	308628	Light A4*	2-6	GTW 1552 p.I-18	1898-1902
		10-1917	21x26"	S	PV	out	73"	180#	24032	SCH	312856	Heavy A4*	7,8	CNR 1556 p.I-16	1904
		3-1923	21x26"	S	PV	out	73"	180#	24032	ROB	312856	Heavy A4*	5,8		1901-1904
A5		not used by GTR													
A6	40	1-1915	21x26"	S	PV	out	73"	175#	23363	SCH	300140		10-13	GTW 1600 p.I-29	1906-1908
A7	10	6-1915	21x26"	S	PV	out	56"	170#	29069	SCH	298052		9		1905
A8	3	1-1915	22x26"	S	PV	out	69"	180#	27904	SCH	308628	Light A8	2,4,7	CNR 1545 p.I-16	1899-1904
A8*	1	4-1920	21x26"	S	PV	out	69"	180#	25425	SCH	312856	Heavy A8	7,8		1904
A9		not used by GTR													
A10	5	6-1919	22x26"	Y	PV	out	69"	200#	31000	SCH	345578	Light A10	5,7,8	GTW 1579 p.I-19	1901-1904
A10+	6	12-1919	22x26"	Y	PV	out	69"	200#	31000	SCH	357114	Heavy A10+	5,6,8	CNR 1584 p.I-21	1901-1904
		1-1923	22x26"	Y	PV	out	69"	200#	31000	ROB	357114	Heavy A10+	5,6,8		1901-1904
A10*	1	10-1920	21x26"	Y	PV	out	69"	200#	31000	SCH	345578	Light A10*	5		1901
		Fitted with these specifications between 10-1920 and 9-1921													
A10†	1	12-1921	21x26"	Y	PV	out	69"	200#	31000	SCH	357114	Heavy A10†	8		1904
		Fitted with these specifications between 12-1921(?) and 3-1922													

Codes:	S = Stephenson	Y = Young valve gear	D = D-valves	PV = Piston valves
	ins = inside ports	out = outside steam pipes		

I-6-a

CNR 1543 (second)   4-6-0 TEN WHEEL TYPE   second I-6-a

		Specifications					Appliances		Weights	Fuel Capacity		Length	Notes
Cylinder	Gear	Driv.	Pressure	Boiler	T.E.	Haulage	Steam	Stkr.	Drivers/Eng./Total	Water	Coal		
20x26"	S	73"	200#	EWT	25425		sat		133/178/308628	5000 gals	10 tons	63-1'	[orig]
20x26"	S	73"	200#	EWT	24219		sat		133/178/308628	5000 gals	10 tons	63-1'	[GTR A]
22x26"	S	69"	180#	EWT	27904	28%	SCH		133/178/308628	5000 gals	10 tons	63-1'	[GTR A8, CNR]

Baldwin Locomotive Works – Burnham, Williams & Company		1898	$11,460							(1) Acquired by CNR 3-01-1923	
	Serial	Shipped	New as	1900	To	7-1908	6-1909	1-1910	— Superheated —	Disposition	To
			—	—	A	A	A	A	(A4) — A8		
1543/2	15688	1- -98	GTR 992*	GTW 992	11-04	GTR 992	nB	GTR 300	1-15 HQ — 1-15 HQ	Sc 7-15-32 LM	
—	15689	1- -98	GTR 993	GTW 993	11-04	GTR 993		GTR 301			CNR 1547
—	15690	1- -98	GTR 994	GTW 994	11-04			GTW 302			GTW 1548
—	15691	1- -98	GTR 995	GTW 995	11-04			GTW 303			GTW 1549

CNR 1543 (second) and 1547-1549 (Lot 2) were built for the **Grand Trunk Railway of Canada** for its lines in the USA. Records in 1905 show GTR 302 and 303 (**GTW 1548** and **1549**) were assigned to the "Western Division" (the *Grand Trunk Western Railway* after 1910). When 1543 and 1547 (as GTR 992 and 993) were transferred, the duty entering Canada amounted to $1400 each. When 1543 was superheated in January 1915, it was modified to the specifications of the A4 class, but when the diameter of the drivers was reduced to 69 inches, it was reclassified into a new class, the A8. By 1923, the four had been modified with open steel cabs. Both **1548** and **1549** were sold to the GTW in January 1924.

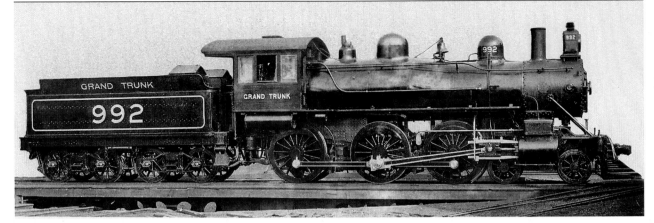

The four Baldwins in GTR's first batch of high-driver 4-6-0s eventually worked both sides of the border in two separate classes. **GTR 992** (1543), at Philadelphia in January 1898, retained its 69-inch drivers and remained in Canada as a GTR A8 and CNR I-6-a.
[BLW PHOTO 1030/H.L. BROADBELT/WES DENGATE COLLECTION]

GTR 993 (1547), at an unidentified location between 1898 and 1910, also stayed in Canada. However, with an increase of driver diameter to 73 inches when superheated, it was assigned to the GTR A4 and CNR I-6-b class respectively.
[AL PATERSON COLLECTION]

Similarly rebuilt as an A4 and I-6-b, **GTW 1549**, at Pontiac, Michigan in the early 1930s, was one of two from the initial order assigned to the United States.
[GEORGE CARPENTER COLLECTION]

CNR 1544-1545 (second)   4-6-0 TEN WHEEL TYPE   second I-6-a

Cylinder	Gear	Driv.	Specifications Pressure	Boiler	T.E.	Haulage	Appliances Steam	Stkr.	Weights Drivers/Eng./Total	Fuel Capacity Water	Coal	Length	Notes
20x26"	S	73"	200#	EWT	24219		sat		133/178/308628	5000 gals	10 tons	63-1'	[orig, GTR A]
22x26"	S	69"	180#	EWT	27904	28%	SCH		133/178/308628	5000 gals	10 tons	63-1'	[GTR A8, CNR]

Grand Trunk Railway – Pointe St. Charles   1899 $9840   (2) Acquired by CNR 3-01-1923

	Serial	Shipped	New as	To	1-1910	Superheated	Steel		Disposition	To
			—	A	A	A8	Hopper			
1544/2	1289	4- -99	**GTR 986***	11-04	**GTR 308**	8-18 HQ	10-22		Sc 5-30-32 JD	
—	1290	4- -99	GTR 987	11-04	GTR 309					**CNR 1556**
—	1291	4- -99	GTR 988	11-04	GTR 310					**CNR 1557**
—	1292	4- -99	GTR 989	11-04	GTR 311					**CNR 1558**
—	1293	6- -99	GTR 990	11-04	GTR 312					**CNR 1559**
1545/2	1294	6- -99	**GTR 991**	11-04	**GTR 313**	12-17 HQ			Sc 11-01-38 MQ	

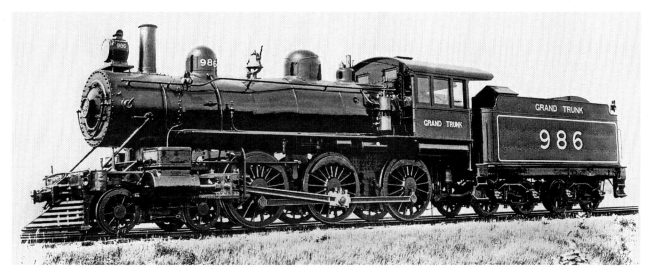

CNR 1544 and 1545 (second) and 1556-1559 (Lot 4) ordered in April 1898, were the first 4-6-0s built by the **Grand Trunk Railway of Canada**. In 1903, GTR paid $2656 each to import **988** and **990** into the USA, to allow them to haul cross-border trains such as the *International Limited*, which had been inaugurated on May 25th 1900. Although the *CNR Locomotive Data Card* for **1559** listed GTR 314 with A4* specs, the entries for both date and shops location for the A4* conversion was blank. CNR I-6-a *Diagram Sheets* give the boiler number for **1545** as #1355, which suggests a possible boiler swap with 1564. By 1923, they had open steel cabs applied.

The third lot of passenger 4-6-0s was built in GTR's Pointe St. Charles shops. Of the six, only two were fitted with smaller diameter drivers. **GTR 986** (1544), at Pointe St. Charles in April 1899, was eventually classed by GTR to A, then A8, and finally by the CNR to I-6-a.
[GTR PHOTO/GEORGE CARPENTER COLLECTION]

The other A8, **GTR 313** (1545), at Turcot between 1910 and 1917, renumbered but not superheated, already had the air pump moved forward to provide easier access to the cab and boiler feed pipe rearranged.
[KARL E. SCHLACHTER PHOTO/GEORGE CARPENTER COLLECTION]

Close inspection of the space between the driving wheels and running board of two locomotives from the same builder's lot reveals the basic difference between the I-6-a and I-6-b classes. **CNR 1545**, at Brantford on June 22nd 1934, had been modified with 69-inch drivers, [AL PATERSON COLLECTION]

whereas **1556**, at Spadina on July 16th 1928, had been rebuilt by the GTR with 73-inch drivers. [H.L. GOLDSMITH/GEORGE CARPENTER COLLECTION] Other alterations made during the six years between the two photographs include the type of pilot, location of classification lights and an extension to the coal bunker.

CNR 1546 (second)						4-6-0 TEN WHEEL TYPE							second I-6-a
	Specifications						Appliances		Weights	Fuel Capacity		Length	Notes
Cylinder	Gear	Driv.	Pressure	Boiler	T.E.	Haulage	Steam	Stkr.	Drivers/Eng./Total	Water	Coal		
21x26"	S	69"	180#	EWT	25425		sat		137/182/312856	5000 gals	10 tons	63-1'	[GTR A8*]

Locomotive & Machine Company of Montreal		1904	(Q-4)	$15,585					(1) Acquired by CNR 3-01-1923	
	Serial	Shipped	New as	To	1-1910	Boiler	— Superheated —		Disposition	
			—	A	A	to	A8	A8*		
1546/2	29856	8- -04	**GTR 957**	11-04	**GTR 339**	324 10-12	11-17 HQ	4-20 HQ	Sc 12-03-35 LM	

CNR 1546: (Lot 7) See **CNR 1572-1577** (page I-22).

CNR 1547 (first);
GTW 1548-1549 (first) **4-6-0** TEN WHEEL TYPE **I-6-b**

Cylinder	Gear	Driv.	Specifications Pressure	Boiler	T.E.	Haulage	Appliances Steam	Stkr.	Weights Drivers/Eng./Total	Fuel Capacity Water	Coal	Length	Notes
20x26"	S	73"	200#	EWT	25425		sat		133/178/308628	5000 gals	10 tons	63-1'	[orig]
20x26"	S	73"	200#	EWT	24219		sat		133/178/308628	5000 gals	10 tons	63-1'	[GTR A]
22x26"	S	73"	180#	EWT	26375		SCH		133/178/308628	5000 gals	10 tons	63-1'	[GTR A4]
21x26"	S	73"	180#	EWT	24032	24%	SCH		133/178/308628	5000 gals	10 tons	63-1'	[GTR A4*, CNR]

Baldwin Locomotive Works – Burnham, Williams & Company   1898 $11,450   (3) Acquired by CNR 3-01-1923

	Serial	Shipped	New	1900	To	7-1908	1-1910	—— Superheated ——		New	Disposition
			—	—	A	A	A	A4	A4*	frame	
1547	15689	1- -98	*GTR* 993	*GTW* 993	11-04	*GTR* 993	GTR 301	2-17 MP	2-19 MP	1-23	Sc 6-30-32 LM
1548	15690	1- -98	*GTR* 994	*GTW* 994	11-04		*GTW* 302		11-17 UB		Sc 11-08-34 UB
1549	15691	1- -98	*GTR* 995	*GTW* 995	11-04		*GTW* 303		9-18 UB		Sc 6-07-35 UB

CNR and **GTW** 1547-1549: (Lot 2) See **CNR** 1543 (second) (page I-13).

GTW 1550-1553 (first) **4-6-0** TEN WHEEL TYPE **I-6-b**

Cylinder	Gear	Driv.	Specifications Pressure	Boiler	T.E.	Haulage	Appliances Steam	Stkr.	Weights Drivers/Eng./Total	Fuel Capacity Water	Coal	Length	Notes
20x26"	S	73"	200#	EWT	24219		sat		133/178/308628	5000 gals	10 tons	63-1'	[GTR A]
22x26"	S	73"	180#	EWT	26375		SCH		133/178/308628	5000 gals	10 tons	63-9½'	[GTR A4]
21x26"	S	73"	180#	EWT	24032		SCH		133/178/308628	5000 gals	10 tons	67-6½'	[GTR A4] ■
21x26"	S	73"	180#	EWT	24032	24%	SCH		133/178/308628	5000 gals	10 tons	63-9½'	[GTR A4*, CNR]

Schenectady Locomotive Works   1898 $11,450   (4) Acquired by CNR 3-01-1923

	Serial	Shipped	New	1900	Rblt	To	1-1910	—— Superheated ——		Disposition
			—	—	nB	A	A	A4	A4*	
1550	4659	4- -98	*GTR* 996	*GTW* 996	-02	11-04	*GTW* 304		1-19 UB ■	Sc 10-31-34 UB
1551/1	4660	4- -98	*GTR* 997	*GTW* 997		11-04	*GTW* 305	6-16 UB	1-18 UB	Sc 11-08-34 UB
1552	4661	4- -98	*GTR* 998	*GTW* 998		11-04	*GTW* 306	5-15 UB	1-18 UB	Sc 5-25-39 UB
1553/1	4662	4- -98	*GTR* 999*	*GTW* 999		11-04	*GTW* 307	7-15 UB	8-18 UB	Sc 5-25-39 UB

GTW 1550-1553 (Lot 3) were built for the **Grand Trunk Railway of Canada**'s "Western Division" and, after 1900, to GTR's reorganized *Grand Trunk Western Railway*. GTR 304 (**1550**) acquired a longer tender, likely when it was superheated. By 1923, all had been rebuilt with open steel cabs. All four were sold to the GTW in January 1924.

GTR 999 (1553), at Schenectady in April 1898, was the last of four built in the second lot of GTR's passenger Ten Wheelers.
[SCHENECTADY WORKS PHOTO S-31/ GEORGE CARPENTER COLLECTION]

Although GTR sequenced the road numbers to follow the first group built by Baldwin (992-995), retention of their 73-inch drivers led CNR to assign them to the I-6-b class, thus breaking the GTR numbering sequence. In Chicago's

Dearborn Station by July 1926, **GTW 1552** had been superheated, modified with a steel cab, and centred headlight. The earlier tender had been replaced or extensively rebuilt.
[C.A. BUTCHER/WES DENGATE COLLECTION]

GTW 1554-1555 (first) — 4-6-0 TEN WHEEL TYPE — I-6-b

Cylinder	Gear	Driv.	Pressure	Boiler	T.E.	Haulage	Steam	Stkr.	Drivers/Eng./Total	Water	Coal	Length	Notes
			Specifications				Appliances		Weights	Fuel Capacity			
20x26"	S	73"	210#	EWT	24219		sat		137/182/312856	7000 gals	10 tons	63-1'	[GTR A]
20x26"	S	73"	200#	EWT	24219		sat		137/182/312856	5000 gals	10 tons	63-1'	[GTR A, 2-1910]
22x26"	S	73"	180#	EWT	26375		SCH		137/182/312856	5000 gals	10 tons	63-1'	[GTR A4]
21x26"	S	73"	180#	EWT	24032	24%	SCH		137/182/312856	5000 gals	10 tons	63-1'	[GTR A4*, CNR]

	Serial	Shipped	New as	To	1-1910	By 1913	— Superheated —		Disposition	To
Schenectady Locomotive Works – ALCO 1904 (S-203) $14,748									(2) Acquired by CNR 3-01-1923	
					A	A	A4	A4*		
			—		A	A	A4	A4*		
1554	29630	6- -04	GTR 964	11-04	GTR 331	GTW 331	3-16 UB	11-17 UB	Sc 11-30-34 UB	
1555	29631	6- -04	GTR 965	11-04	GTR 332	GTW 332	12-15 UB	11-17 UB	Sc 11-30-34 UB	
—	29632	6- -04	GTW 966	11-04	GTW 333					GTW 1579
—	29633	6- -04	GTW 967*	11-04	GTW 334					GTW 1580
—	29634	6- -04	GTW 968	11-04	GTW 335					GTW 1581

GTW 1554, 1555 and **1579-1581** (Lot 7) were built for **Grand Trunk Railway of Canada**. Records show GTR 331 and 332 (1554 and 1555) had been reassigned by 1913 to the *Grand Trunk Western Railway*. GTR 333-335 (1579-1581) had each been superheated to A10* lightweight specifications with Young valve gear by the USRA, GTR and *GTW*, and numbered into CNR's I-7-a class. By 1923 they all had been rebuilt with open steel cabs. All five were sold to the GTW in January 1924. Tenders of both **1580** and **1581** were rebuilt with cast steel underframes.

The high-driver Ten Wheelers from GTR's seventh order were to spend most of their careers stateside. **GTR 967** (1580), at Schenectady in June 1904, was built with 73-inch drivers and Stephenson valve gear.
[SCHENECTADY WORKS PHOTO S-301/AL PATERSON COLLECTION]

Both **GTW 1581**, at Battle Creek in 1938,
[KENNETH S. MacDONALD/WES DENGATE COLLECTION]
and **GTW 1579**, also at Battle Creek in July 1940,
[HAROLD K. VOLLRATH/DON McQUEEN COLLECTION]

were rebuilt with 69-inch drivers and Young valve gear.
This change in specification resulted in being classed by
CNR to I-7-a. The first pair in the order retained their as-built
specifications and became members of the I-6-b class.

CNR 1556-1559 (first)   4-6-0 TEN WHEEL TYPE   I-6-b

Cylinder	Gear	Driv.	Pressure	Boiler	T.E.	Haulage	Steam	Stkr.	Drivers/Eng./Total	Water	Coal	Length	Notes
			Specifications				Appliances		Weights	Fuel Capacity		Length	Notes
20x26"	S	73"	200#	EWT	24219		sat		133/178/308628	5000 gals	10 tons	63-1'	[GTR A]
22x26"	S	73"	180#	EWT	26375		SCH		133/178/308628	5000 gals	10 tons	63-1'	[GTR A4]
21x26"	S	73"	180#	EWT	24032		SCH		133/178/308628	5000 gals	10 tons	63-1'	[GTR A4*]

Grand Trunk Railway – Pointe St. Charles		1899	$9840							(4) Acquired by CNR 3-01-1923	
	Serial	Shipped	New as	5-1903	To	1-1910	Boiler	—— Superheated ——		Disposition	
			—		A	A	from	A4	A4*		
1556	1290	4- -99	**GTR 987**		11-04	**GTR 309**			6-19 MP	Sc 11-30-31 JD	
1557	1291	4- -99	**GTR 988**	*GTR 988*	11-04	**GTR 310**			11-18 MP	Sc 7-15-32 LM	
1558	1292	4- -99	**GTR 989**		11-04	**GTR 311**		10-17 HQ		Sc 6-20-35 MV	
1559	1293	6- -99	**GTR 990**	*GTR 990*	11-04	**GTR 312**	339 10-12	7-16 MP (11-20)?		Sc 6-3-32 LM	

CNR 1556-1559: (Lot 4) See **CNR 1544-1545** (page I-15).

CNR 1560-1565 (first) 4-6-0 TEN WHEEL TYPE I-6-b

Cylinder	Gear	Driv.	Pressure	Boiler	T.E.	Haulage	Steam	Stkr.	Drivers/Eng./Total	Water	Coal	Length	Notes
			Specifications				Appliances		Weights	Fuel Capacity		Length	Notes
20x26"	S	73"	200#	EWT	24219		sat		133/178/308628	5000 gals	10 tons	63-1'	[GTR A]
22½&35x26"	S	73"	200#	EWT	24219?		sat		133/ / 000	5000 gals	10 tons	63-1'	[GTR A 4-6-0C]
22x26"	S	73"	180#	EWT	26375		SCH		133/178/308628	5000 gals	10 tons	63-9½'	[GTR A4]
21x26"	S	73"	180#	EWT	24032		SCH		133/178/308628	5000 gals	10 tons	67-6½'	[GTR A4] ■
21x26"	S	73"	180#	EWT	24032	24%	SCH, ROB		133/178/308628	5000 gals	10 tons	63-9½'	[GTR A4*]

Grand Trunk Railway – Pointe St. Charles 1901 $9760 (6) Acquired by CNR 3-01-1923

	Serial	Shipped	New as	2-1903	To	3-1905	1-1910	A	A4	A4*	Mods	Disposition	To
			—	—	A		A						
1560/1	1349	-01	**GTR 978**	—	11-04		**GTR 314**			R 6-23 HQ	nC nF	Sc 6-30-32 LM	
1561	1350	-01	**GTR 979**		11-04		**GTR 315**	5-15 HQ				Sc 12-20-35 HW	
—	1351	-01	GTR 980		11-04		GTR 316						**CNR 1578**
1562	1352	-01	**GTR 981**	4-6-0C	11-04	4-6-0	**GTR 317**			S 3-18 MP ■		Sc 10-22-31 LM	
1563	1353	-01	**GTR 982**		11-04		**GTR 318**	8-14 HQ		■	nC nF	Sc 10-31-31 JD	
—	1354	-01	GTR 983		11-04		GTR 319						**CNR 1583**
1564/1	1355	-01	**GTR 984**		11-04		**GTR 320***	11-17 HQ	S 4-28 MP			Sc 10-19-31 LM	
1565/1	1356	-01	**GTR 985**		11-04		**GTR 321**		S 8-23 HQ			Sc 11-30-31 JD	
—	1357	-01	CVR 220	GTR 969	11-04		GTR 322						**CNR 1584**
—	1358	-01	CVR 221	GTR 970	11-04		GTR 323	2-15 HQ					**CNR 1566**
—	1359	-01	CVR 222	GTR 971	11-04		GTR 324	9-15 HQ	S (12-21)?				**CNR 1567**
—	1360	-01	CVR 223	GTR 972	11-04		GTR 325						**CNR 1585**

CNR 1560-1567, 1578 and 1583-1585 (Lot 5) were twelve locomotives built in two lots during 1901 by the **Grand Trunk Railway of Canada**. GTR 981 (1562) was rebuilt in 1903 as a Richmond compound, remaining in service as such for two years. GTR 322-325 (1584, 1566, 1567, 1585) were leased to the **Central Vermont Railway** sometime in 1912 and were returned before being superheated. In December 1912, the boiler from GTR 312 (1559) was applied to GTR 339 (1546) and that of 339 cascaded to GTR 324 (1567). GTR 316 (1578) was superheated with a new boiler into GTR's A10 lightweight class with Young valve gear. CNR later numbered it into the I-7-a class. It carried 21x26-inch cylinders for nine months between 1920 and 1921. GTR 319 (1583)

During 1901 and 1902, GTR built seventeen more passenger 4-6-0s with 72-inch drivers. Initially four in this fourth lot of passenger Ten Wheelers were built for the CVR (see CVR 220-223, page CV-29). **GTR 317** (1562) at Bonaventure Station, Montreal sometime after 1910, experimentally compounded between 1903 and 1905, had yet to be superheated or have any alterations made to either its cylinder size or driver diameters.
[AL PATERSON COLLECTION]

was superheated into GTR's A10 heavyweight class with Young valve gear and eventually went into the CNR I-7-b class. Although the *CNR Locomotive Data Card* for **1567** listed GTR 324 with A4* specs, the entries for both

(text continues on next page)

date and shops location for the A4* conversion was blank. GTR 317-318 (**1562** and **1563**) acquired longer tenders, likely when they were superheated. New steel cabs (nC) were installed on **1560** and **1563** (as GTR 314, 318) in December 1921. As CNR locomotives, both received new frames (nF), a "new design" for **1560** in June 1923, and a "heavier" one for **1563** in January 1926. By 1923, all had been modified with open steel cabs. After retirement, both **1584** and **1585** were stored in the Scarboro pit.

GTR 319 (1583), at St. Albans, Vermont during September 1914, had yet to be superheated with Young valve gear as a GTR A10. When 1563 (see Vol. I, p. 102) was rebuilt with 22 x 26-inch cylinders and 73-inch drivers, it was assigned the I-6-b class. [KARL E. SCHLACHTER PHOTO/GEORGE CARPENTER COLLECTION]

Built as CVR 220, **1584**, at Turcot on October 11th 1937, had been rebuilt into the heavyweight GTR A10 class with Young gear, thus becoming a CNR I-7-b. Both, by the time of the 1930s photographs, had been altered to CNR style practice with the addition of running board ladders and steel cabs. Sometime during the decade, 1584 was fitted with a raised coal bunker. [AL PATERSON COLLECTION]

I-6-b

CNR 1566-1567 (first) — 4-6-0 TEN WHEEL TYPE — I-6-b

Cylinder	Gear	Driv.	Pressure	Boiler	T.E.	Haulage	Steam	Stkr.	Drivers/Eng./Total	Water	Coal	Length	Notes
			Specifications				Appliances		Weights	Fuel Capacity			
20x26"	S	73"	200#	EWT	24219		sat		133/178/308628	5000 gals	10 tons	63-1'	[GTR A]
22x26"	S	73"	180#	EWT	26375		SCH		133/178/308628	5000 gals	10 tons	63-1'	[GTR A4]
21x26"	S	73"	180#	EWT	24032		SCH, ROB		133/178/308628	5000 gals	10 tons	63-1'	[GTR A4*]

Grand Trunk Railway – Pointe St. Charles   1901 $10,145   (2) Acquired by CNR 3-01-1923

	Serial	Shipped	New as	2-1903	To	1-1910	CVR lease	Superheated		Mods	Disposition	To
			F	—	A	A	c.1912 > c.1915	A4	A4*			
—	1357	-01	CVR 220	GTR 969	11-04	GTR 322	CVR 220					CNR 1584
1566	1358	-01	CVR 221	GTR 970	11-04	GTR 323	CVR 221	2-15 HQ		Sc 11-23-35 HW		
1567	1359	-01	CVR 222	GTR 971	11-04	GTR 324	CVR 222	9-15 HQ	(12-21)?	Sc 3-28-36 MQ		
—	1360	-01	CVR 223	GTR 972	11-04	GTR 325	CVR 223					CNR 1585

CNR 1566 and 1567: (Lot 5) See CNR 1560-1565 (page I-20) and CVR 220-223 (page CV-29).

CNR 1568-1571 (first)   4-6-0 TEN WHEEL TYPE   I-6-b

		Specifications					Appliances		Weights	Fuel Capacity		Length	Notes
Cylinder	Gear	Driv.	Pressure	Boiler	T.E.	Haulage	Steam	Stkr.	Drivers/Eng./Total	Water	Coal		
20x26"	S	73"	200#	EWT	24219		sat		133/178/308628	5000 gals	10 tons	63-1'	[GTR A, pv]
22x26"	S	73"	180#	EWT	26375		SCH		133/178/308628	5000 gals	10 tons	63-9½'	[GTR A4]
21x26"	S	73"	180#	EWT	24032		SCH		133/178/308628	5000 gals	10 tons	67-6½'	[GTR A4] ■
21x26"	S	73"	180#	EWT	24032	24%	SCH		133/178/308628	5000 gals	10 tons	63-9½'	[GTR A4*]

Grand Trunk Railway – Pointe St. Charles   1902 $11,550   (4) Acquired by CNR 3-01-1923

	Serial	Shipped	New as	Rblt	To	1-1910	New	New	——— Superheated ———		Disposition	To
				nB & pv	A	A	Cyls	Boiler	A4	A4*		
—		1361	-02	GTR 973	-04	11-04	GTR 326						**CNR 1586**
1568	1362	-02	**GTR 974**		11-04	**GTR 327**			9-15 HQ	(12-21)?	Sc 6-30-32 LM	
1569	1363	-02	**GTR 975**	-04	11-04	**GTR 328**				1-18 MP	Sc 7-15-32 LM	
1570	1364	-02	**GTR 976**		11-04	**GTR 329**			3-15 HQ? 11-16 HQ	1-20 MP	Sc 7-15-32 LM	
1571	1365	-02	**GTR 977**	-06	11-04	**GTR 330**	2-14 MP		8-19 MP ■	(12-21)?	Sc 7-15-32 LM	

CNR **1568-1571** and **1586** (Lot 6) were ordered in March 1902 and built simple by the **Grand Trunk Railway of Canada**. They were GTR's first locomotives rebuilt with piston valves (pv). Although the *CNR Locomotive Data Card* for **1559** listed GTR 314 with A4* specs, the entries for both date and shops location for the A4* conversion was blank. One, GTR 326 (**CNR 1586/1**), was superheated into GTR's A10 heavyweight class with Young valve gear and later went into the CNR I-7-b class. GTR 330 (1571) acquired a longer tender, likely when it was superheated. By 1923, all had been modified with open steel cabs.

CNR 1572-1577 (first)   4-6-0 TEN WHEEL TYPE   I-6-b

		Specifications					Appliances		Weights	Fuel Capacity		Length	Notes
Cylinder	Gear	Driv.	Pressure	Boiler	T.E.	Haulage	Steam	Stkr.	Drivers/Eng./Total	Water	Coal		
20x26"	S	73"	225#	EWT	24219		sat		137/182/317770	7000 gals	10 tons	63-1'	[orig, GTR A]
20x26"	S	73"	210#	EWT	24219		sat		137/182/312856	5000 gals	10 tons	63-1'	[GTR A, 6-1907]
20x26"	S	73"	200#	EWT	24219		sat		137/182/312856	5000 gals	10 tons	63-1'	[GTR A, 2-1910]
22x26"	S	69"	180#	EWT	26375		SCH		137/182/312856	5000 gals	10 tons	63-1'	[GTR A4]
21x26"	S	73"	180#	EWT	24032	24%	SCH, ROB		137/182/312856	5000 gals	10 tons	63-1'	[GTR A4*, CNR]

Locomotive & Machine Company of Montreal   1904-1905   (Q-4) $15,585   (6) Acquired by CNR 3-01-1923

	Serial	Shipped	New as	To	1-1910	——— Superheated ———			Disposition	To
			—	A	A	A4	A4*	I6b		
—	29853	8- -04	GTR 954	11-04	GTR 336					**CNR 1587**
1572	29854	8- -04	**GTR 955**	11-04	**GTR 337**		S 1-19 MP		Sc 4-29-36 LM	
—	29855	8- -04	GTR 956	11-04	GTR 338					**CNR 1588**
—	29856	8- -04	GTR 957	11-04	GTR 339					**CNR 1546/2**
1573	29857	8- -04	**GTR 958***	11-04	**GTR 340**		S 12-24 MP		Sc 2-28-36 MQ	
1574/1	29858	1- -05	**GTR 959**	1-05	**GTR 341**			R 3-23 MP	Sc 2-20-36 MQ	
1575	29859	1- -05	**GTR 960**	1-05	**GTR 342**		S 6-18 MP		Sc 10-01-35 LM	
1576/1	29860	1- -05	**GTR 961**	1-05	**GTR 343**	S 7-15 HQ	7-19 MP		Sc 3-31-36 LM	
1577	29861	1- -05	**GTR 962**	1-05	**GTR 344**		S 10-17 MP		Sc 1-07-36 MQ	
—	29862	12- -04	GTR 963	12-04	GTR 345					**CNR 1582**

The eighth lot of Ten Wheelers was eventually split into three CNR classes. **GTR 959** (1574), at an unidentified location in western Ontario between 1905 and 1910, was to retain its as-delivered specifications and become an I-6-b class steamer.
[J.J. TALMAN REGIONAL COLLECTION, D.B. WELDON LIBRARY, U.W.O.]

I-6-b

GTR 344 (1577), also at an unidentified location, but after 1910, with a classic Grand Trunk front end, had been renumbered but not superheated. It too, retained its as-built specifications. [DICK GEORGE/WES DENGATE COLLECTION]

CNR 1572-1577, 1546, 1582, 1587 and **1588** (Lot 8) were built for the **Grand Trunk Railway of Canada.** When acquired, GTR 340 (**1573**) had yet to be superheated by GTR, and the work was done by the CNR. By 1923, they had open steel cabs applied. GTR 345 (**1582**) was superheated into GTR's A10* lightweight class with Young valve gear, and was later numbered into the CNR I-7-a class. GTR 336 and 338 (**1587** and **1588**) were superheated into GTR's A10 heavyweight class with Young valve

(text continues on next page)

Three in the last order of 4-6-0s with 73-inch driving wheels were eventually rebuilt to lightweight A10 (I-7-a) and heavyweight A10 (I-7-b) specifications. Another was modified with 69-inch drivers and classed as an A8 (I-6-a). Both photographed at Spadina, I-6-b **1573** in the early 1930s, [DICK GEORGE/WES DENGATE COLLECTION] superheated to A4 specifications by CNR, and **1577**, on July 16th 1928 (page I-24), exhibited some interesting differences. [H.L. GOLDSMITH/GEORGE CARPENTER COLLECTION] Although both had centred headlights (changed by the GTR), **1573** had already been shopped with a boiler tube pilot, running board ladders, relocated class lamps and air reservoir, and had its forward side windows blanked.

I-6-b

gear and later went into the CNR I-7-b class. Although the *CNR Locomotive Data Card* for **1587** listed GTR 336 with A10 specs, the entries for both date and shops location for the change to a 21-inch stroke was blank.

See caption on page I-23.

CNR, GTW 1578-1588 (first)   4-6-0 TEN WHEEL TYPE   I-7-a; I-7-b

GTR stock books broke the A10 class into either light or heavyweight 4-6-0s. The heavier-weighted Ten Wheelers are marked in the roster as A10+. Rationale for two sets of weights has never been satisfactorily explained, although one explanation suggests different GTR statisticians may

have made stock books entries based on light (empty) engine and tender weights for some, and loaded for others. In 1923, CNR assumed the data significant enough to group the locomotives into two I-7 sub-classes as 1578-1582 (-a) and 1583-1588 (-b).

CNR 1578 (first)   4-6-0 TEN WHEEL TYPE   I-7-a

Specifications							Appliances		Weights	Fuel Capacity		Length	Notes
Cylinder	Gear	Driv.	Pressure	Boiler	T.E.	Haulage	Steam	Stkr.	Drivers/Eng./Total	Water	Coal		
20x26"	S	73"	200#	EWT	24219		sat		133/178/308628	5000 gals	10 tons	63-1'	[orig, GTR A]
22x26"	Y	69"	200#	EWT	31000		SCH		148/198/345578	6250 gals	10 tons	66-0'	[GTR A10, CNR]
21x26"	Y	69"	200#	EWT	31000	31%	SCH		148/198/345578	6250 gals	10 tons	66-0'	[GTR A10*]

Grand Trunk Railway – Pointe St. Charles 1901 $9760										(1) Acquired by CNR 3-01-1923	
	Serial	Shipped	New as	11-1904	1-1910	Superheated				Stl	Disposition
						& new boiler	A10*	A10		hopper	
						22x26	21x26	22x26			
1578	1351	-01	GTR 980	GTR 980	GTR 316	6-19 MP	10-20 MP	9-21 MP		5-22	Sc 10-19-31 LM

CNR 1578: (Lot 5) See **CNR 1560-1565** (page I-20).

GTW 1579-1581 (first)   4-6-0 TEN WHEEL TYPE   I-7-a

Specifications							Appliances		Weights	Fuel Capacity		Length	Notes
Cylinder	Gear	Driv.	Pressure	Boiler	T.E.	Haulage	Steam	Stkr.	Drivers/Eng./Total	Water	Coal		
20x26"	S	73"	210#	EWT	24219		sat		137/182/312856	5000 gals	10 tons	63-1'	[orig, GTR A]
20x26"	S	73"	200#	EWT	24219		sat		137/182/312856	5000 gals	10 tons	63-1'	[GTR A, 2-1910]
22x26"	Y	69"	200#	EWT	31000	31%	SCH		148/198/345578	6250 gals	10 tons	66-0'	[GTR A10, CNR]

Schenectady Locomotive Works – ALCO 1904 (S-203) $14,748									(3) Acquired by CNR 3-01-1923	
	Serial	Shipped	New as	11-1904	1-1910	——— Superheated ———		Cast		Disposition
						A10	I7a	tender of		
1579	29632	6- -04	GTW 966	GTW 966	GTW 333	2-20 UB (USRA)				Sc 7-07-45 UB
1580	29633	6- -04	GTW 967*	GTW 967	GTW 334	10-21 UB		6-27 UB		Sc 7-07-45 UB
1581	29634	6- -04	GTW 968	GTW 968	GTW 335	12-23 UB	5-26 UB			Sc 7-07-45 UB

I-6-b

I-7-a

GTW 1579-1581: (Lot 7) See **GTW 1554-1555** (page I-18); the boiler pressure note under **CNR 1572-1577** (page I-23); and the GTR A10 class note under **CNR 1578-1588** (page I-24).

CNR 1582 (first) — 4-6-0 TEN WHEEL TYPE — I-7-a

Cylinder	Gear	Driv.	Specifications Pressure	Boiler	T.E.	Haulage	Appliances Steam	Stkr.	Weights Drivers/Eng./Total	Fuel Capacity Water	Coal	Length	Notes
20x26"	S	73"	225#	EWT	24219		sat		137/182/317770	7000 gals	10 tons	63-1'	[orig, GTR A]
20x26"	S	73"	210#	EWT	24219		sat		137/182/312856	5000 gals	10 tons	63-1'	[GTR A, 6-1907]
20x26"	S	73"	200#	EWT	24219		sat		137/182/312856	5000 gals	10 tons	63-1'	[GTR A, 2-1910]
22x26"	Y	69"	200#	EWT	31000	31%	SCH		148/198/345578	6250 gals	10 tons	66-0'	[GTR A10, CNR]

Locomotive & Machine Company of Montreal		1904	(Q-4)	$15,585				(1) Acquired by CNR 3-01-1923
	Serial	Shipped	New as	11-1904	1-1910	Superheated		Disposition
			—	A	A	A10		
1582	29862	12- -04	GTR 963	GTR 963	GTR 345	10-19 HQ		Sc 10-12-34 LM

CNR 1582: (Lot 8) See **CNR 1572-1577** (page I-22) and the GTR class note for A10 under **CNR 1578-1588** (page I-24).

CNR 1583-1586 (first) — 4-6-0 TEN WHEEL TYPE — I-7-b

Cylinder	Gear	Driv.	Specifications Pressure	Boiler	T.E.	Haulage	Appliances Steam	Stkr.	Weights Drivers/Eng./Total	Fuel Capacity Water	Coal	Length	Notes
20x26"	S	73"	200#	EWT	24219		sat		133/178/308628	5000 gals	10 tons	63-1'	[orig, GTR A; pv]
22x26"	Y	69"	200#	EWT	31000	31%	SCH, ROB		156/210/357114	6250 gals	10 tons	66-0'	[GTR A10+, CNR]

Grand Trunk Railway – Pointe St. Charles		1901-1902	$:various (see below)							(4) Acquired by CNR 3-01-1923
	Serial	Shipped	New as	New as	Orig cost	6-1902	To	1-1910	CVR Lease	Superheated	Disposition
			F	—			A	A	c-12>-15	& nB A10+	
1583	1354	12- -01		GTR 983	$ 9,760			11-04	GTR 319		R 1-23 HQ	Sc 12-03-35 LM
1584	1357	-01	CVR 220		$10,145	GTR 969	11-04	GTR 322	CVR 220	S 5-21 HQ	Sc 8-29-39 JD
1585	1360	12- -01	CVR 223		$10,145	GTR 972	11-04	GTR 325	CVR 223	S 4-20 HQ	Sc 9-23-38 MV
1586/1	1361	1- -02		GTR 973	$11,550			11-04	GTR 326		S 9-20 HQ	Sc 12-03-35 LM

CNR 1583-1585: (Lot 5) See **CNR 1560-1567** (pages I-20 and I-21) and CVR 220-223 (page CV-29).

1586/1: (Lot 6) See **CNR 1568-1571** (page I-22).

CNR 1587-1588 (first) — 4-6-0 TEN WHEEL TYPE — I-7-b

Cylinder	Gear	Driv.	Specifications Pressure	Boiler	T.E.	Haulage	Appliances Steam	Stkr.	Weights Drivers/Eng./Total	Fuel Capacity Water	Coal	Length	Notes
20x26"	S	73"	225#	EWT	24219		sat		137/182/317770	7000 gals	10 tons	63-1'	[orig, GTR A]
20x26"	S	73"	210#	EWT	24219		sat		137/182/312856	5000 gals	10 tons	63-1'	[GTR A, 6-1907]
20x26"	S	73"	200#	EWT	24219		sat		137/182/312856	5000 gals	10 tons	63-1'	[GTR A, 2-1910]
22x26"	Y	69"	200#	EWT	31000	31%	SCH		156/210/357114	6250 gals	10 tons	66-0'	[GTR A10+, CNR]
21x26"	Y	69"	200#	EWT	31000	31%	SCH		156/210/357114	6250 gals	10 tons	66-0'	[GTR A10+, 1921?-22]

Locomotive & Machine Company of Montreal		1904	(Q-4)	$15,585					(2) Acquired by CNR 3-01-1923
	Serial	Shipped	New as	To	1-1910	New	Superheated	21x26"	22x26"	Disposition
			—	A	A	boiler	A10+ 22x26"	A10†	A10+	
1587	29853	8- -04	GTR 954	11-04	GTR 336		12-19 HQ	(12-21)?	3-22 MP	Sc 3-31-36 LM
1588	29855	8- -04	GTR 956	11-04	GTR 338	9-21 HQ	9-21 HQ			Sc 10-12-35 LM

CNR 1587 and 1588: (Lot 8) See **CNR 1572-1577** (page I-22).

CNR 1589-1598 (first)   4-6-0 TEN WHEEL TYPE   I-8-a

Specifications							Appliances		Weights	Fuel Capacity		Length	Notes
Cylinder	Gear	Driv.	Pressure	Boiler	T.E.	Haulage	Steam	Stkr.	Drivers/Eng./Total	Water	Coal		
19x26"	S	73"	210#	EWT	21858		sat		126/167/286000	6000 gals	12 tons	61-8'	[GTR A3]
19x26"	S	73"	200#	EWT	21858	22%	sat		126/167/300140	5000 gals	10 tons	61-8'	[GTR A3 2-1910; CNR]
21x26"	S	73"	175#	EWT	23363	23%	sat		126/167/300140	5000 gals	10 tons	64-5'	[GTR A6] ■
21x26"	S	73"	175#	EWT	23363	23%	SCH		126/167/300140	5000 gals	10 tons	61-8'	[GTR A6, CNR]

Locomotive & Machine Company of Montreal 1906 (Q-33) $16,558									(10) Acquired by CNR 3-01-1923
	Serial	Shipped	New as	1-1910	New	—— Superheated ——		New	Tender	Disposition
			A3	A3	Boiler	A6	I8a	stl cab	to	
1589	39538	6- -06	GTR 1000	GTR 400		2-18 HQ				Sc 12-30-35 MQ
1590	39539	6- -06	GTR 1001	GTR 401	7-14 MP		8-24 MP	12-21		Sc 12-03-35 LM
1591	39540	7- -06	GTR 1002	GTR 402		8-18 HQ				Sc 12-30-35 MQ
1592	39541	7- -06	GTR 1003	GTR 403	11-19 HQ	11-19 HQ			OCS	Sc 9-19-38 MQ
1593	39542	7- -06	GTR 1004/2	GTR 404		10-16 MP ■				Sc 1-04-36 LM
1594	39543	7- -06	GTR 1005/2	GTR 405		1-17 HQ				Sc 4-08-36 MQ
1595	39544	7- -06	GTR 1006/2	GTR 406		11-20 HQ		12-21		Sc 4-29-31 JD
1596	39545	7- -06	GTR 1007/2	GTR 407		1-16 HQ				Sc 12-23-35 LM
1597	39546	7- -06	GTR 1008/2	GTR 408		7-22 MP				Sc 12-03-35 LM
1598	39547	7- -06	GTR 1009/2*	GTR 409		6-20 HQ		12-21		Sc 12-28-35 HW

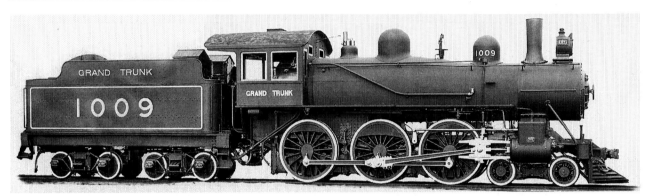

CNR 1589-1598 (Lot 10), ordered in November 1905, were built for the **Grand Trunk Railway of Canada**. GTR 404 (1593) acquired a longer tender, likely when superheated. When acquired by CNR, only GTR 401 (**CNR 1590**) remained to be superheated. By 1923, the class had been modified with open steel cabs. The tender of **1592** was set aside in 1938 to be converted to an OCS water transport. After retirement, **1593** was temporarily stored in the Scarboro pit.

Both the GTR and CNR allocated the passenger Ten Wheelers built between 1906 and 1908 into a class different from those built earlier, even though they had 73-inch drivers. These later 4-6-0s which ultimately became the CNR I-8-a class were lighter in weight, and had smaller cylinder dimensions. **GTR 1009** (1598), at Montreal in June 1906, was one of the first ten built.
[MLW PHOTO Q-24/DON McQUEEN COLLECTION]

Renumbered in 1910 and rebuilt in 1914 (note the cab and tender), **GTR 401** (1590), at Palmerston in May 1923, had yet to be superheated.
[C.A. BUTCHER/WES DENGATE COLLECTION]

Steaming away in the late afternoon sun, **1591**, at Spadina terminal in Toronto about 1933, had acquired the standard CNR front end but had yet to have the forward side windows blanked. The extended smokebox apparent on all three examples (GTR 1009, GTR 401 and CNR 1591) remained unaltered throughout their thirty years of service. [AL PATERSON COLLECTION]

CNR 1500-1516 (second)	(17) See CNR **1111=1160** G-16-a class (page G-18)	
CNR 1517-1519 (second)	(3) See CNR **1161=1165** G-17-a class (page G-21)	
CNR 1520 (second)	(1) See CNR **1221=1230** H-4-a class (page H-9)	
CNR 1521 (second)	(1) See CNR **1268=1277** H-6-b class (pages H-15 & H-16)	
CNR 1522-1531 (second)	(10) See CNR **1278=1322** H-6-c class (pages H-16, H-18 & H-19)	
CNR 1532-1545 (second/third)	(14) See CNR **1323=1342** H-6-d class (page H-22)	
CNR 1546-1550 (second)	(5) See CNR **1347=1351** H-6-f class (page H-26)	
CNR 1551-1589 (second)	(39) See CNR **1354=1409** H-6-g class (pages H-29 & H-31)	
CNR 1590-1594 (second)	(5) See CNR **1423=1442** H-10-a class (page H-36)	
CNR 1595-1599 (second)	(5) See CNR **1443=1452** H-10-a class (page H-38)	

I-8-a

GTW 1599-1600 (first);
GT   1601-1603;
CNR 1604-1608   4-6-0 TEN WHEEL TYPE   I-8-a

Cylinder	Gear	Driv.	Specifications Pressure	Boiler	T.E.	Haulage	Appliances Steam	Stkr.	Weights Drivers/Eng./Total	Fuel Capacity Water	Coal	Length	Notes
19x26"	S	73"	210#	EWT	21858		sat		126/167/286000	5000 gals	10 tons	61-8'	[GTR A3]
19x26"	S	73"	200#	EWT	21858		sat		126/167/286000	5000 gals	10 tons	61-8'	[GTR A3 2-1910]
21x26"	S	73"	175#	EWT	23363		sat		126/167/300140	6000 gals	10 tons	64-5'	[GTR A6] ■
21x26"	S	73"	175#	EWT	23363	23%	SCH		126/167/300140	6000 gals	10 tons	61-11⅛'	[GTR A6, CNR]

Schenectady Locomotive Works – ALCO   1906   (S-383) $14,617   (10) Acquired by CNR 3-01-1923

	Serial	Shipped	New as A3	1-1910 A3	1912-16	Superheated A6	To I-8-a	To I-8-a	Tender to	Disposition
1599	40623	8-30-06	**GTR 1010**	*GTW* 410		10-18 UB	**GTW** -23			Sc 9- -43 UB
1600	40624	8- -06	**GTR 1011**/2	*GTW* 411		4-18 UB	**GTW** -23			Sc 5-10-35 UB
1601	40625	8- -06	**GTR 1012**/2	*GTR/GT* 412		1-16 UB	**GT** 12-14-23			Sc 12-07-35 DJ
1602	40626	8- -06	**GTR 1013**	*GTR/GT* 413		9-16 HQ	**GT** 4-11-25		OCS	Sc 11-02-38 DJ
1603	40627	8- -06	**GTR 1014***	*GTR/GT* 414		8-15 UB & nB	**GT** 6- -23	**CNR** 9-30-23		Sc 12-28-35 HW
1604	40628	8- -06	**GTR 1015**	*GTR/GT* 415		10-16 UB	(GT)			Sc 12-20-35 HW
1605	40629	8- -06	**GTR 1016**	*GTR* 416		6-20 MP			OCS	Sc 11-15-41 LM
1606	40630	8- -06	**GTR 1017**	*GTR* 417		2-20 MP				Sc 10-15-38 MQ
1607	40631	8- -06	**GTR 1018**/1	*GTR* 418	CVR Ls	7-15 HQ			OCS	Sc 10-01-38 LM
1608	40632	8-31-06	**GTR 1019**/2	*GTR* 419	**CVR** 419	10-19 HQ & nB ■				Sc 9-04-41 LM

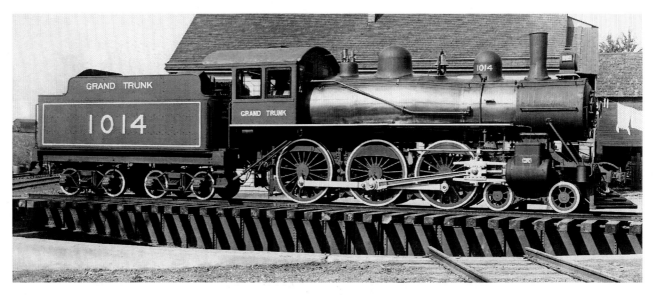

CNR 1599-1600, 1604-1608 and GT 1601-1603 (Lot 11) were all built either for **Grand Trunk Railway of Canada** or its US lines. Records indicate GTR 410 and 411 (**1599** and **1600**) remained assigned to the *Grand Trunk Western Railway* and were lettered GTW in 1923 and after the formation of the **Grand Trunk Western Railroad** in 1928. GTR 412-414 were assigned by GTR to its eastern US lines at the time of CNR amalgamation. Two were permanently relettered "Grand Trunk" (**1601-1602**) and remained on the eastern division. GTR 414 was briefly lettered **GT 1603** but within months was reassigned to Canada and relettered CNR. Although CNR records show GTR 415 (**1604**) assigned to Portland, Maine as of January 30th 1923, it was transferred to Richmond, Quebec during July 1923 and never was lettered "Grand Trunk" **1604**, even though the 4-6-0 briefly worked the New England line in August 1924. Saturated GTR 418 and 419 (**1607** and **1608**) were leased to Central Vermont

The second group of future I-8-a 4-6-0s came from the United States. Undeterred by the closeness of the builder's turntable, the proverbial Monday mornings wash hung on a line behind **GTR 1014** (**1603**) at Schenectady in August 1906. The early morning light illuminated all the detail of this well-proportioned Grand Trunk passenger engine. [SCHENECTADY WORKS PHOTO S-383/DON McQUEEN COLLECTION]

between 1912 and 1916, but only GTR 419 became **CVR 419** during the lease.

GTR 419 (**1608**) acquired a longer tender, likely when it was superheated. Two received new boilers (nB) at same time they were superheated. By 1923, the class had been equipped with open steel cabs. In December 1926, CNR paid $2296 in duties and taxes in order to allow **1608** to operate exclusively in Canada. After retirement, **1607** was stored in the Scarboro pit before being scrapped at London. The tenders of **1602**, **1605** and **1607** were set aside to be converted to OCS water transports.

Remaining well groomed, but renumbered to **GTR 414** (1603), at South Paris, Maine during July 1915, it appeared to have undergone few, if any, modifications. During the next five years, however, major technical changes would permanently alter this pristine appearance. The cable in the foreground, threaded through a series of white posts (see also GTR 438 on page I-31), was to a semaphore ahead of the locomotive.
[C.A. BUTCHER/WES DENGATE COLLECTION]

Five from the Schenectady order were to remain in the United States. Superheated **GTW 1600** at Detroit's Brush Street Station on March 5th 1934 had been modified with the usual smokebox front adornments and blanked cab windows, and shopped with a standard GTW flat metal bar pilot and stirrup running board steps. **GT 1602**, at Portland, Maine on December 12th 1936, had been similarly modified, except for Canadian shops applying the preferred running board ladder and boiler tube pilot.
[BOTH: AL PATERSON COLLECTION]

I-8-a

CNR 1609-1618   4-6-0 TEN WHEEL TYPE   I-8-a

			Specifications				Appliances		Weights	Fuel Capacity		Length	Notes
Cylinder	Gear	Driv.	Pressure	Boiler	T.E.	Haulage	Steam	Stkr.	Drivers/Eng./Total	Water	Coal		
19x26"	S	73"	210#	EWT	21858		sat		126/167/286000	5000 gals	10 tons	61-8'	[GTR A3]
19x26"	S	73"	200#	EWT	21858		sat		126/167/300140	5000 gals	10 tons	61-8'	[GTR A3 2-1910]
21x26"	S	73"	175#	EWT	23363		sat		126/167/300140	6000 gals	10 tons	64-5'	[GTR A6] ■
21x26"	S	73"	175#	EWT	23363	23%	SCH/ROB		126/167/300140	6000 gals	10 tons	61-8'	[GTR A6, CNR]

Grand Trunk Railway – Pointe St. Charles   1907 $10,303   (10) Acquired by CNR 3-01-1923

	Serial	Shipped	New as A3	1-1910 A3	Superheated A6	I8a	Stl hopper	New stl cab	Tender to	Disposition
1609	1480	-07	GTR 1020/3	GTR 420	R 3-18 HQ		3-20			Sc 7-41-41 MQ
1610	1481	-07	GTR 1021/3	GTR 421	S 5-15 HQ					Sc 10-09-35 LM
1611	1482	-07	GTR 1022/2	GTR 422		S 9-24 MP				Sc 3-31-36 LM
1612	1483	-07	GTR 1023/2	GTR 423	■ R 2-23 HQ				OCS	Sc 9-27-38 LM
1613	1484	-07	GTR 1024/3	GTR 424	S 5-20 HQ			12-21		Sc 1-15-36 HW
1614	1485	-07	GTR 1025/2	GTR 425	S 4-15 HQ					Sc 4-25-36 LM
1615	1486	-07	GTR 1026/2	GTR 426	S 5-15 HQ					Sc 7-15-32 LM
1616	1487	-07	GTR 1027/2	GTR 427	S 10-20 MP				OCS	Sc 10-01-38 LM
1617	1488	-07	GTR 1028/2	GTR 428		S 9-23 MP		12-21		Sc 6-23-39 LM
1618	1489	-07	GTR 1029/2	GTR 429	S 12-20 MP					Sc 1-04-36 LM

CNR 1609-1618 (Lot 12) were the last 4-6-0s built by the **Grand Trunk Railway of Canada**. GTR 423 (1612) acquired a longer tender, likely when it was superheated. By 1923, all of the group had been modified with open steel cabs. **CNR 1609** was stored in the Scarboro pit between 1937 and 1940, before it was dismantled. The tenders of **1612** and **1616** were set aside in 1938 for conversion to OCS water transports, but were deemed unsuitable and scrapped in September 1939.

Grand Trunk built the third batch of future I-8-a Ten Wheelers in 1907. At the end of its career, **1609** was at the Scarboro pit in May 1937. During those three decades, GTR effected the usual modifications, including superheating and headlight centring, and after 1923, CNR added a boiler tube pilot, steel cab and coal bunker extension. The GTR bell location between the domes was rarely moved forward during any CNR shopping. The large-diametered drivers were the reason why the I-8 classes acquired the moniker "Liners", because of their less-than-desirable habit of "hunting" or yawing back and forth between the rails at high speeds. [AL PATERSON/WES DENGATE COLLECTION]

I-8-a

GTW 1619-1626;
CNR 1627-1628

4-6-0 TEN WHEEL TYPE

I-8-a

Cylinder	Gear	Driv.	Pressure	Boiler	T.E.	Haulage	Steam	Stkr.	Drivers/Eng./Total	Water	Coal	Length	Notes
			Specifications				Appliances		Weights	Fuel Capacity			
19x26"	S	73"	210#	EWT	21858		sat		126/167/288000	6000 gals	10 tons	61-8'	[GTR A3 orig]
19x26"	S	73"	200#	EWT	21858		sat		126/167/288000	5000 gals	10 tons	61-8'	[GTR A3 2-1910]
21x26"	S	73"	175#	EWT	23363	23%	SCH		126/167/300140	5000 gals	10 tons	61-8'	[GTR A6, CNR]

Baldwin Locomotive Works – Burnham, Williams & Company   1908 $16,188   (10) Acquired by CNR 3-01-1923

	Serial	Shipped	New as A3	1-1910 A3	Superheated A6		Disposition
1619	32774	5- -08	**GTR 1030**/2	**GTW 430**	1-15 UB	Rs 12-31-34	Sc 7-29-41 UB
1620	32775	5- -08	**GTR 1031**/2*	**GTW 431**	12-15 UB		Sc 7-07-45 UB
1621	32776	5- -08	**GTR 1032**/2	**GTW 432**	8-20 UB		Sc 4-12-35 UB
1622	32777	5- -08	**GTR 1033**/2	**GTW 433**	8-17 UB		Sc 8- -43 UB
1623	32799	5- -08	**GTR 1034**/2	**GTW 434**	4-16 UB		Sc 11-08-34 UB
1624	32803	6- -08	**GTR 1035**/2	**GTW 435**	1-18 UB		Sc 12-15-34 UB
1625	32808	6- -08	**GTR 1036**/2	**GTW 436**	9-19 UB		Sc 10-30-44 UB
1626	32809	6- -08	**GTR 1037**/2	**GTW 437**	1-19 UB & nX		Sc 5-31-35 UB
1627	32810	6- -08	**GTR 1038**/2	**GTW 438** GTR 438 by 6-10	6-16 UB		Sc 4-06-36 MQ
1628	32811	5- -08	**GTR 1039**/2	**GTW 439** GTR 439 by 6-10	4-16 HQ		Sc 3-23-36 MQ

Baldwin built the fourth and last batch of what was to become the I-8-a class. The enhanced illustration, made from a photograph of saturated **GTR 1031** (1620), at Philadelphia in May 1908, [BLW LITHOGRAPH 2676/H.L. BROADBELT/WES DENGATE COLLECTION] clearly showed the extended smokebox, slatted pilot and location of the right-hand appliances. Renumbered 1038 as **GTR 438** (1627), likely at South Paris, Maine during July 1915, had yet to undergo the front-end changes brought about by superheating (see 1619 and 1625 on

page I-32.) With the removal of the saturated cylinders and the steam chests, whose top had doubled as a foothold, the running boards were extended to the smokebox front and fitted with either steps, stirrups or ladders. The former location of the classification lamps had been predetermined by their accessibility at the end of the running boards, but with the extension of the running boards above the superheated cylinder castings, the lights were brought forward to the smokebox face. [H.L. GOLDSMITH/GEORGE CARPENTER COLLECTION]

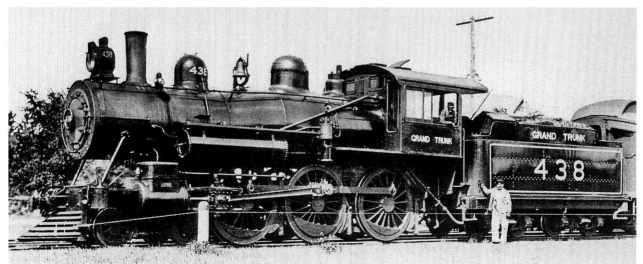

GTW 1619-1626 and CNR 1627-1628 (Lot 13) were ordered in December 1907, five for the **Grand Trunk Railway of Canada** and five for the *Grand Trunk Western Railway*. Records show GTR 430-434 (1619-1623) were assigned to the USA by 1910. All but two were lettered GTW in 1923, and after the formation of the **Grand Trunk Western Railroad** in 1928. By 1923, they all had been rebuilt with open steel cabs. GTW 1624-1626 were sold to the GTW in January 1924. In December 1926, CNR paid $2433 each in duties and taxes in order to allow **1627** and **1628** to operate exclusively in Canada. **GTW 1619** was removed from company records on December 31st 1934 but was reinstated on January 31st 1935, and replaced by **1621**.

Only two of the Baldwin order were to remain in Canada, while the other eight were eventually assigned Grand Trunk's US *western lines*. GTW 1619, at Durand on January 17th 1937, was one of four in the lot to remain in service until the mid-1940s. [AL PATERSON COLLECTION]
Another survivor, GTW 1625, at Durand on April 8th 1939, was the second-last to be retired. [GEORGE CARPENTER COLLECTION]
Both were very typical GTW in appearance, although there were some differences. GTW 1625 had retained spoked pilot wheels and acquired a larger extension to its coal bunker. The location of metal classification flags varied; either beside the headlight, as in the case of 1619, or on top of the smokebox, as found on 1625. The road number lamp of 1619 was mounted in a higher position than that on 1625. In later years, headlights on many GTW steamers appeared to be off-centre, a phenomenon created when larger-diameter headlights were replaced with more compact models, without any change being made to the bracket location.

1629-1699   Numbers not used